Back to the
Astronomy Café

Back to the Astronomy Café

More Questions and Answers about the Cosmos from "Ask the Astronomer"

Sten Odenwald

Westview
PRESS

A Member of the Perseus Books Group

Copyright © 2003 by Westview Press, A Member of the Perseus Books Group

Published in the United States of America by Westview Press, A Member of the Perseus Books Group, 5500 Central Avenue, Boulder, Colorado 80301-2877, and in the United Kingdom by Westview Press, 12 Hid's Copse Road, Cumnor Hill, Oxford OX2 9JJ.

Find us on the world wide web at www.westviewpress.com

Westview Press books are available at special discounts for bulk purchases in the United States by corporations, institutions, and other organizations. For more information, please contact the Special Markets Department at the Perseus Books Group, 11 Cambridge Center, Cambridge, MA 02142, or call (617) 252-5298, (800) 255-1514 or email j.mccrary@perseusbooks.com.

A Cataloging-in-Publication data record for this book is available from the Library of Congress.
ISBN 0-8133-4166-3

The paper used in this publication meets the requirements of the American National Standard for Permanence of Paper for Printed Library Materials Z39.48-1984.

Text design by Trish Wilkinson
Typeset in 11-point Adobe Garamond by the Perseus Books Group

10   9   8   7   6   5   4   3   2   1

This book is dedicated to the astronauts of Columbia who gave their lives so that the rest of us might hold on to our dreams of exploring space.

Michael Anderson
David Brown
Kalpana Chawla
Laurel Clark
Rick Husband
William McCool
Ilan Ramon

They were born on Earth, yet their spirits touched the stars.

Contents

Introduction

The last five years have taken us on what can only be described as an emotional, astronomical roller-coaster ride. It began with the breathtaking announcements of new planets discovered by the dozens—mysterious worlds known only by their masses, distances, and orbit periods. It has ended with the convincing discovery of dark energy, a presidential initiative for nuclear propulsion development, and the deaths of seven astronauts. No one can really say where these events will take us in the years to come. Some people worry about the human costs. They question why so many of our precious resources are being squandered when machines can do the exploration job more cheaply. Some say that, instead of finding answers to silly questions about space, we should be spending more money on social concerns. The more than $2 trillion we spend each year on the U.S. government doesn't seem to be quite enough to cure poverty and make us completely safe. Others of us focus on the successes of the human exploration of cosmic mysteries. We see how this great adventure and accumulation of knowledge have quietly, but profoundly, transformed our entire civilization without any fanfare. Most people can name famous athletes, politicians, and movie stars, but cannot name the inventor of the microwave, the computer, the transistor, or very likely a single physicist involved in the quantum revolution. Even professional scientists would come up short in such a test. Those who choose to quibble about the "horrific" cost of space exploration need only look into their children's eyes at the mention of space travel or black holes. It is a glimmer of hope and excitement about what the future might bring that you will not find if you lower their eyes to the ruts and furrows of daily life.

As we survey the new landscape of knowledge we have fashioned for ourselves, we can easily find just cause to admire our creativity and technological prowess. We should also feel proud of how much we have learned in our journey. In 100 years, scientists have virtually closed the books on many cosmic questions, and found answers to a thousand more that no one ever thought to ask. In the last five years we have found ourselves thrust into a dark cosmos, yet it is also a cosmos where every shining star sports a discoverable retinue of worlds to explore. The universe may contain inscrutable dark energies and matter, but at least in our little starry corner of it we have found familiar things to wonder about. The rest of the universe with its distant galaxies is an unreachable phantom world. Only a few more details remain to explore before this realm can be put aside as merely a mute background for human existence within the Milky Way. Exploring neighboring planetary systems, however, opens up a new and exciting window for us to look through. The adrenaline begins to mount, and we dare to ask once again that timeless question "Is there anyone else out there?"

It isn't in the murky images of infant galaxies that humans will finally come to know this answer but in the steadily mounting tally of planets detected around nearby stars. The process, once begun, will be relentless. First we will know how commonplace Jupiter-sized worlds are. Then we will count up the planetary systems with circular orbits. Next, we will learn about the abundance of Earth-sized worlds as their minute tugs become detectable in our data. From there, we will telescopically probe thousands of planetary atmospheres searching for oxygen and chlorophyll. Newspapers will eventually report the discovery of the first extraterrestrial biosphere, which many of us astronomers are convinced must surely exist somewhere out there. At that moment in history, the concept of life in the cosmos will become a calculable quantity like the mass of the universe in dark matter, or the luminosity of the Sun. On the back of an envelope, we will be able to estimate the total biomass of the Milky Way. Artists and poets will swoon over the prospects of new worlds to paint and alien sunsets to describe. And what will we do then? To know that, orbiting a nameless star 197 light-years away, an Earth-sized world exists with a biosphere will evoke only one response: We have to go there—somehow. We may have to spend a trillion dollars and work for a century, but some extension of humanity or its silicon technology will eventually reach this distant planet and report back. Will we find a world covered by bacteria, dinosaurs, or intelligence? We can only dream.

This is my fourth book in a series that began with *The Astronomy Café* (Freeman, 1998), *The 23rd Cycle: Learning to Live with a Stormy Star* (Columbia University Press, 2000), and *Patterns in the Void: Why Nothing Is Important* (Westview, 2002). My previous three books turned out to be a trilogy without my conscious intent to write them that way. *The Astronomy Café* explored numerous short questions of the kind for which an astronomer would design a research program to find specific answers. This is an important and very basic way that scientists operate on a day-to-day basis as they create new knowledge. This is the level of scientific research that the general public most understands, and is the reason why my *Ask the Astronomer* web site has been so successful over the years. Then came the higher-order process of science represented by *The 23rd Cycle*, where numerous lines of investigation and data gathering are threaded together to create a theory for why things happen the way they do. This theory focuses on a bigger picture that tries to forecast what will happen next. It is here that science research really shines and becomes extremely practical. We forecast weather, predict the orbits of planets and their motion in the night sky, and trace the future destiny of our sun as a star. But there is something missing in these two scientific enterprises. Where do humans fit into this great tapestry of cause-and-effect? In *Patterns in the Void,* I tried to answer this kind of question by revealing where scientific knowledge can take you if you are willing to make the journey. It is a powerfully emotional trip through astronomy and physics at the highest levels of abstraction, and emotional contemplation, in search of that elusive concept called "meaning." The journey, at least for this astronomer, was not always pleasant, and in the end the conclusions reached seemed to remain aloof from the effort expended in finding them.

So why do I once again find myself at the beginning, writing what appears to be a sequel to the first book in my original trilogy? Perhaps, unconsciously, it is an attempt to plunge back into the world of fresh data and new advances, and to begin the journey again to the climax of a second trilogy. Given the opportunity, will I find myself writing the sequel to *Patterns in the Void* with a completely new outlook and conclusion? What a marvelous goal to look forward to!

I am so grateful to my family for their staunch support of my writing. My wife, Susan, has always helped me get through the tedious and painful process of giving birth to a new book. My daughters, Emily and Stacia, are natural explorers of the unknown, and have even helped me overcome my fear of the dark—except at Halloween! I would like to thank Holly Hodder, Vice-President of Westview Press, for her encouragement and help in editing my trilogy, and suggesting I return to the Astronomy Café. My editor at Westview, Steve Catalano, has been a great help in reading through the manuscript and helping me keep my answers focused. Astronomers do love to ramble on about the minutiae of a subject, especially those that are dear to their hearts! I would especially like to thank Rebecca Marks, the Senior Project Editor for this book at Westview, for all of her work in the final editing of the manuscript and galley proofs. Finally, and most importantly, I want to thank the literally hundreds of thousands of visitors to my web resources who have found these questions and answers so interesting over the years. The Astronomy Café (www.astronomycafe.net) has been, for me, a wonderful opportunity to meet fellow cosmic travelers. I only wish I could offer real coffee to my visitors! Although many of you prefaced your E-mails with "I'm sorry to ask you such a dumb question but . . . ," I rarely found any of them to be dumb or silly. If you think you have a dumb question, you are probably thinking like a scientist because many of our questions seem pretty bone-headed, too!

CHAPTER 1

The World's Greatest Hits

Since 1997, when *The Astronomy Café* web site began its long tenure on the Internet, millions of visitors have passed through its portals looking for a few simple answers to their most pressing questions about the cosmos. During the six years since then, some questions of the thousands that were answered and posted quickly distinguished themselves as the all-time favorites. It was simply a matter for me to collect them all together, update them as needed, and offer them to an even wider community in printed form. The questions you will find in this chapter are both endearing in their simplicity and impressive in their universality. Some are humorous while others are extremely deep and specialized. Some are ages-old while others are as new and fresh as a walk through a contemporary art museum. All are, however, a fascinating snapshot of what we are thinking as we begin to inhabit the 21st century. My electronic tally shows that each has been asked by more than 30,000 people since the Astronomy Café first opened its doors. Enjoy!

◑ ● ◐

1   Is Earth about to reverse its magnetic field?

It sure looks that way, but don't hold your breath. It will take several thousand years for us to get to the point where this could actually happen. The last reversal ended about 780,000 years ago, and we now live in what geologists call the Brunhes Chron, but there have been 171 of these reversals of magnetic polarity during the last 78 million years. Careful studies of the magnetism in rocks show us that Earth—like a car engine left out in the cold—has lots of trouble "turning over." Data obtained from the Ocean Drilling Program by Steve Lund at the University of Southern California and his colleagues in 2001 paint this picture of magnetic activity as it is recorded in sediment deposits. Since the last major polarity reversal that gave us our current magnetic

1

conditions, there have been 14 times when the field has weakened and strengthened without reversing its polarity, an event called a sub-chron or an "excursion." Figure 1 gives you some hint of what has happened to the field in the last 8,000 years. What would a reversal look like when it is in full swing? It helps to know that there are actually two different magnetic fields to Earth.

There is the major "bar magnet" field called the dipole component, and then there is the crustal field. The crustal field was created when molten rock solidified and remembered how the local dipole field was oriented. Over time, and millions of years of evolution, the crust has become a patchwork of local fields that add together to equal about 10 percent of the strength of Earth's main dipole field. Geologists use this crustal field to identify oil and ore-bearing regions that register as slight magnetic changes as instruments scan across them. This field more or less stays put and moves around very slowly as the continents move, or as new crust is produced. The main "bar magnet" field, which helps us navigate by compass, is created by currents in the liquid outer core of Earth. Its strength increases and decreases, and the locations of its north and south poles move around on the surface, depending on how these currents flow. As the orientation of these currents changes within Earth, the main field also moves around on the surface. This is what causes cartographers to remake their harbor maps every few decades to keep up with the changing magnetic bearings for "true north."

During the last few hundred years, scientists have had equipment sensitive enough to track Earth's field accurately. The intensity of the main field seems

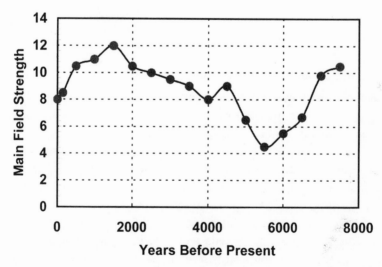

FIGURE 1 Earth's field reversal during the last 8,000 years. (CREDIT: AUTHOR)

to be falling by about 5 percent per century. This means that if the rate stays constant, Earth's main field will be no stronger than the irregular crustal field in about 1,000 or 2,000 years. But is it really going to be a steady decline? We don't know because we have never been able to study other reversals with year-by-year resolution. The latest Canadian Arctic Survey in 2001 reported an accelerating change in the motion of the pole. After staying near the area of northwest Canada for nearly a century, it will be moving rapidly to stand nearly directly over the North Pole by about 2030, and then by the 22nd century it will be in northern Siberia! Because the magnetic field is a critical buffer, shielding the atmosphere from cosmic rays and charged particles from the Sun, you might think that the health hazard posed by Earth losing its magnetic field is rather fearful to think about. The geologic record, however, shows that Earth's polarity flips every half-million years or so and dips to a strength near the level of the crustal field. There are no identifiable fossil effects from previous reversals, so perhaps they are not as much of a problem for the biosphere as we might suppose. We know that many species of bacteria, birds, and fish can sense the direction of the magnetic field and use it for migration and finding food in murky waters. These animals, or the less adaptable members of a species, may perish "suddenly" when the field changes and they can no longer rely on it to orient themselves. We have never experienced losing our main field. The fossil record is pretty mute about the biological effects. By the way, in March 2003, an awful movie called *The Core* showed all hell breaking loose as the field began to vanish. The biggest reason not to expect anything to happen this time is that, 780,000 years ago, when the last field reversal happened, it was a complete washout. No grand extinction events. No evidence for atmospheric chaos. Nothing.

2   Will the Earth ever stop spinning?

The probability for such an event is practically zero in the next few billion years. But suppose something like this could happen? The outcome depends on just how suddenly it stopped. If the Earth stopped spinning very quickly, say within a day, the atmosphere would still be in motion with the Earth's original 1,100-miles-per-hour rotation speed at the equator. All of the landmasses would be scoured clean of anything not attached to bedrock. If the process happened gradually over billions of years, the situation would be very different, and it is this possibility that is the most likely as the constant torquing of the Sun and Moon upon the Earth finally reaches its conclusion. As for other effects, presumably the magnetic field of the Earth is generated by a dynamo effect that involves its rotation. If the Earth stopped rotating, its magnetic field would no longer be regenerated. It would decay to some low, residual value due to the very small component that is "fossilized" in its crustal rocks. There would be no more "northern lights" and the Van Allen radiation belts would probably vanish, as would our protection from cosmic rays and other high-energy particles. Cosmic rays are a significant biohazard for astronauts in space, but not for those of us living on the ground, thanks to our atmosphere, which shields us almost completely from them.

Will its spin ever change in the future through natural causes? We don't know. The only way we think it could change in any radical way is by a major impact with another celestial body. Even the biggest asteroid we know, Ceres (636 miles in diameter), is not really big enough to do much more than annihilate all life on this planet. It might tilt the Earth's axis by a few degrees depending on the details of the collision. But much of the kinetic energy of the impact would be used in blasting a portion of the outer atmosphere into space along with billions of tons of crustal rock. There are no foreseeable asteroid collisions with Earth big enough to be important in triggering more than some minor surface damage for the foreseeable future. What we can't completely rule out are changes in the way that mass is distributed on the crust. If you move the continents around, some geologists think you could unbalance the spinning Earth and make it flip its rotation axis in space. They even think this may already have happened. It's called the Snowball Earth model. (See Question 37.)

3   Where do the extra four minutes go in a 23-hour 56-minute day?

Earth orbits the Sun once every 365.256 days. This is how long it takes the Sun to travel once around the sky through each of the zodiacal constellations and end up in exactly the same star field it started from. The calendar, however, only has 365 honest-to-God slots. In terms of hours, one rotation of Earth takes 23 hours, 56 minutes, and four seconds. This means that there is 0.256 of a day left over each year, which after four years adds up to 4 × 0.256 = 1.024 days. Every four years, we decide to add a leap day at the end of February, which only has 28 days in a standard year. This still leaves us at the end of four years with 0.024 days or 34.4 minutes extra. This extra bit of unaccounted time in our calendar, which is trying to keep up with the exact motion of the Sun in the sky, is made up in a rather non-intuitive way. Every two centuries we add a second day. 1600 was a leap year, 1700, 1800, and 1900 were not, but in 2000 we had 366 days with February 29th added. This follows the rule that century years divisible by 400 *are* leap years to make up for that extra 0.024 days we still have after four years. The next leap day will be in 2004.

4 How old is the Earth?

Ancient rocks exceeding 3.5 billion years in age are found on all of Earth's continents. The oldest rocks on Earth are the Acasta Gneisses in northwestern Canada near Great Slave Lake (4.03 billion years) and the Isua Supracrustal rocks in west Greenland (3.7 to 3.8 billion years), but well-studied rocks nearly as old are also found in the Minnesota River Valley and northern Michigan (3.5–3.7 billion years), in Swaziland (3.4–3.5 billion years), and in western Australia (3.4–3.6 billion years). Mineral grains (zircon) from sedimentary rocks in west central Australia, which have been radioactively dated using the uranium-lead method, yield ages of 4.4 billion years and are the oldest remnants of Earth's first crust. The oldest dated moon rocks have ages between 4.4 and 4.5 billion years and provide a minimum age for the formation of our nearest planetary neighbor. Thousands of meteorites, which are fragments of asteroids that fall to Earth, have been recovered. The results show that the meteorites, and therefore the solar system, formed between 4.53 and 4.58 billion years ago. A consistent age for the whole she-bang is therefore closer to 4.54 billion years, with an uncertainty of less than 1 percent. The Moon formed about 100 million years later.

5 Is there any substance to the idea that Earth is 6,000 years old?

None whatsoever. Creationists have published many documents that claim to show evidence that Earth is this much younger than the scientific facts allow. It all started with the Scottish Bishop James Ussher (1581–1686), who performed a literal reading of the Bible and a numeration of the ages of the earliest inhabitants. This led to a creation of the universe in the year 4004 B.C., so Earth must be (4004 + 2003) or 6,006 years old. To make this work, modern biblical fundamentalists have to assert that radioactive dating (see Glossary) is a scientific fraud, and that it leads to results in direct conflict with the Word of God. They have collected a variety of discordant age estimates, or references to conflicting results in the literature, but when you look closely, you will find that their references are often 100 to 200 years out of date, or they have completely misrepresented the scientific evidence. The good news is that Creationists are a vast minority within the family of religious people on this planet. In many instances, their fervent beliefs on this matter are in conflict with the leaders in the Protestant and Catholic churches from which they claim to derive their authority. Still, I suspect we will be hearing about Creationism for at least another thousand years, because there is no social, political, or religious penalty to be paid for harboring these kinds of opinions.

6   Where does the atmosphere end and outer space begin?

The definition depends very distinctly on the technology. Astronauts get their space "wings" when they fly above 50 miles. Stratospheric research balloons can ascend to 30 miles (161,000 feet). High-altitude manned balloons ascend to 25 miles (132,000 feet). Each of these altitudes can be considered an "edge to space." Some jet fighters define the "edge of space" as low as 80,000 feet, where you can see the black of the sky above and the blue of the Earth below and its curvature as well. In reality, the atmosphere just gets less and less dense the higher up you go. Even at 7,000 miles, there is a trace of Earth's "hydrogen cloud" at its outer atmosphere, called the geocorona. Beyond this lies the plasmasphere, an even more remote atmospheric outpost extending to 15,000 miles or more depending on solar storm conditions.

7   Were the Apollo moon landings really a hoax?

Of course not, but the concept certainly did sell a lot of books for the authors of this loony idea. Sadly, it also got a lot of people, including a disturbingly large number of teachers and students, worked up about ideas that they simply didn't have the science background to sort out properly. This idea began in 2001 with a Fox-TV program called *Conspiracy Theory: Did We Land on the Moon?* that was aired on February 15 and repeated on March 21. The program interviewed "experts" who strongly believed that NASA never had the technology to make it to the Moon. They claimed that NASA was under political pressure to make us look better than the Soviets at any cost. The issues raised by the "Hoaxers," however, were very deceptive and gave ammunition to people who don't trust scientists, and who think that all phenomena should have an obvious intuitive explanation. Some of my favorite questions have to do with the American flag continuing to wave long after it was planted, the lack of a major crater under the lunar excursion module (LEM), and the seemingly anomalous shadows. You get anomalous shadows because Earth is a very bright source of light at a different angle than the Sun. The American flag continued to wave because of the violence with which it was slammed into the surface and the frictionless environment.

There is no atmosphere on the Moon to damp the motion quickly. Finally, the large rocket nozzle of the engine distributed the thrust of the rocket exhaust over a huge area. This results in just enough momentum change to slow down the LEM, but hardly enough pressure at the ground to move much of the lunar surface around—even the dust.

8   Will a rogue comet or asteroid hit Earth?

Comets are unpredictable. We know the orbits of hundreds of comets, but the problem is that every year, a dozen new comets enter the inner solar system thanks to events that occur way out in the Kuiper Belt beyond the orbit of Saturn. We cannot see bodies out there that are smaller than a few hundred miles across. This means that we can't even see most of the bodies that can damage Earth until they are a few years away from actually colliding with Earth. It can take weeks to get a good orbit determination. In the next 50 years, the known comets will be our friends. It is the hundreds of currently unknown ones discovered by amateur astronomers during this same time that are the most frightening. One of the most spectacular finds by amateurs has been the NASA/ESA Solar and Heliospheric Observatory satellite data. Over 500 previously unknown comets have been discovered in these images.

Asteroids present a different problem. Although there are no known asteroids on a direct collision course, as of March 12, 2003, the International Astronomical Union's Minor Planets Center currently has a list of over 498 Potentially Hazardous Asteroids with closest approaches less than 4.6 million miles. Most of these have been discovered since 1998, when careful surveys for them were started by NASA. They can be as small as a few dozen yards, or as large as a few miles across. With a little bad luck, any one of these could get gravitationally tweaked by Jupiter so that their orbits intersect Earth. If you want to keep up with this ongoing survey, visit the NASA/JPL NEO pages on the web (http://neo.jpl.nasa.gov/neo.html). (See Question 54.)

9   What is a "blue moon"?

According to *Sky and Telescope* magazine, this term has a very complicated and surprisingly recent origin. The expression "once in a blue moon" appears to have a centuries-old history. It is supposed to refer to a very rare event. But its more exact definition as the second full moon in a single month seems to be only a few decades old, and is due to a mistake made in an article published in *Sky and Telescope* back in the 1940's. This definition has only really become popular since the 1980's. The second full moon in a month begs to be called something, and to call it a blue moon seems just as good as any other name I can come up with. There are 29.53 days between any two full moons. There are 365.24 days in the year, which equals 12.37 lunar months. The little bit that is left over (0.37) means that there can be 13 full moons during years that are about $1/0.37 = 2.7$ years apart. There will be blue moons on the 31st day of July 2004, December 2009, August 2012, July 2015, January 2018, March 2018, October 2020, August 2023, May 2026, and December 2028. There will also be a blue moon on June 30, 2007. 1999 had two blue moons, and the next year with two will be 2018.

10   When will NASA develop a true "nuclear" rocket engine?

We already have, in fact several times. All rockets work the same way. They throw things out one end of the spacecraft, and this causes the spacecraft to move in the opposite direction to conserve momentum. A nuclear reactor makes heat, and this can be used to drive a turbine to produce electricity, or to vaporize a fuel and expel it as rocket momentum (nuclear thermal rocket engine). In the former case, the electricity can be used to drive an ion engine, or a mini-magnetosphere propulsion system. In 1959, NASA had a program called NERVA (1955–1961), but it was canceled soon after the 900-megawatt "KIWI" nuclear rocket was test-fired. This was replaced by the Phoebus engine in 1965, which operated at 1,500 megawatts. The most powerful engine of all was the 5,000-megawatt Phoebus 2A tested in June 1968, which ran for 12 minutes at a temperature of 3,700 K. The problem that engineers encountered with all of these engines was the erosion of the nuclear core by the flowing rocket gases. About 20 percent of the core would be ejected out the nozzle every five hours in a highly radioactive plume. Forty years later, Project Prometheus has been funded by Congress to develop a high-yield satellite nuclear power in the next five years at a cost of $3 billion. It is not specifically a plan to design a new rocket engine but to create small reactors to power satellites and planetary rovers in deep space. As Ed Weiler, chief of the Office of Space Science at NASA, puts it, "For 40 years, NASA has been doing planetary science in the same way. That is, you accelerate for 5 to 10 or 15 minutes and then you stop . . . and you coast, and you coast, and you coast. Occasionally, Jupiter will be just in the right spot and you can get a little slingshot effect, but it's not always there. That's not the way to do exploration. That's exploring the West by going in covered wagons." NASA now has the green light to build state-of-the-art nuclear-electric generators that will carry us to the outer solar system or to a new world of sophisticated energy-hungry planetary rovers. They will also be used to generate power for sophisticated high-acceleration ion engines. (See Questions 127 and 338.)

11   What would happen if an astronaut took a glove off in space?

You would expect to get a bad case of frostbite as perspiration instantly turns to ice on your skin. Within a few seconds, the water in the surface skin cells would turn to ice crystals and possibly rupture the cell walls. This wave of cell freezing would move deeper into the body until your hand would be completely frozen. At about the same time, the gas in the blood stream would start to boil off and rupture the skin all over your body. Nerve cells might continue to function for many seconds after exposure, and record intense pain until they too stopped working. But in seconds, the freezing process would continue beyond your hand and engulf the rest of your body exposed to vacuum. Once this happened, a number of physiological changes would occur very quickly, and unconsciousness would set in after about 30 seconds to a minute. It takes time for gas to pass out from the inside of the body and escape into space. There would be lung damage as the alveoli ruptured and fluids and gases flowed into the lung cavities. The result? More intense pain, convulsions, and unconsciousness. However, and incredibly, it is predicted that even after 30 to 40 seconds you could still be resuscitated! But would you want to be?

12   How fast could we travel in space with existing technology?

On October 28, 1998, the Deep Space-1 (DS-1) mission flew the first-ever ion rocket engine as a primary propulsion system for an interplanetary spacecraft. Meanwhile, a twin engine has been running continuously on Earth for 24,750 hours (1.8 years) under the supervision of a team of engineers led by Anita Sengupta at the Jet Propulsion Laboratory. The thrust of this 1,300-watt, 1,200-volt engine (about 0.02 pounds) is microscopic by comparison with the million-pound-thrust engines used to launch the space shuttle. But this little motor can operate continuously and could propel a 25-pound payload to a speed of 18,400 miles per hour after three years, or a whopping 65,000 miles per hour after 10 years. Newer designs using nuclear electricity and not solar cells would easily quadruple this performance and get us to 100,000 miles/hour in only four years or so. This technology is already space-qualified with DS-1 spacecraft. And, oh by the way, the total weight of the propellant for DS–1 was only about 250 pounds! (See Question 338.)

13   How much money has NASA spent on space research?

The recent budgets are at a level of about $15 billion and amount to just under 1 percent of the total U.S. federal budget. In the 1960's this fraction was closer to 5 percent. Since the 1980's, it has been slowly declining until by 2003, it is at 0.7 percent. We spend as much on NASA as we do buying potted plants and gardening materials every year. Economically, it's hard to defend the notion that we, as a nation, have a very serious attitude toward space. About 10 percent of the NASA budget goes to aviation. The remainder goes to space research both manned and unmanned. Since the beginning of NASA back in 1959, a total of $466 billion has been spent by NASA over 44 years when correction to "2003 dollars" is made to take inflation into account. Figure 2 shows how the NASA budget has fared compared to the federal budget, and that it is on a decline. Some of us call it free fall.

It's interesting to note that $466 billion is about equal to a single year's expenditure by the Department of Defense. It is well known that NASA is horribly underfunded in light of the objectives and missions it is asked to carry out. If the budget were doubled, you would see many more very ambitious engineering and scientific projects to detect life beyond Earth, and planets orbiting other stars. There would be a full-fledged lunar research outpost with some very impressive telescopes in operation. During the last 20 years, we have spent endless time debating why we need a space station, why we really need to go to the Moon and Mars, etc. While the debating goes on, and budgets are reduced, we lose precious opportunities to carry out these projects at lower cost than what we now have to pay. When the USSR was still our enemy, the debate was about political prowess and national security. But today, it is much harder to convince anyone to do anything that costs money other than spend for the military, medicine, and social security. It will take a national calamity such as a direct asteroid impact on a major city to change this. The second problem is that our basic understanding of how to carry out large projects is still rudimentary. We do not know how to put someone in space for more than 200 days without serious medical impacts. We do not know how to build a closed biological system for long-term habitation. Our technologies for putting material in orbit still cost thousands of dollars a pound. Even if we had more money, it is not obvious how to accelerate the learning process, but it would sure help to have more resources and people involved in the research.

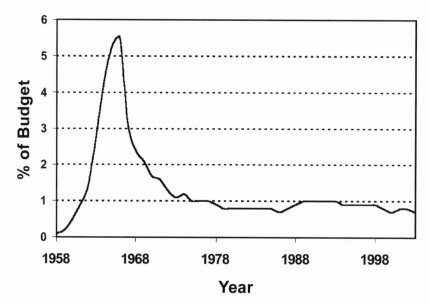

FIGURE 2 The percentage of the federal budget for NASA and space research has not kept pace with our dreams, or with the pace of scientific discovery since the Apollo years. Entire generations weaned on *Star Trek* and *Star Wars* demand more from NASA than what we are financially willing to commit to this great adventure. Because NASA must obey the laws of nature, and most of these laws require expensive solutions, NASA cannot continue this decline without great human and technological risk. (CREDIT: AUTHOR)

14 When do you think humans will set foot on Mars?

The pessimist in me finds it very hard to believe that this will ever happen. For Mars, no country has yet found a reason for going there that can be supported politically or economically. When you tell politicians that we need to go to Mars to simply explore its mysteries and search for fossils, or to continue the great human exploration of new frontiers, they look at you blankly and say, "Yes that's true, but you need a better reason." Plus, there are some very serious technological and medical problems. No one has figured out how to ensure that the astronauts will be strong enough to make it to Mars and back without deteriorating their bones, experiencing radiation sickness, and suffering a range of serious physiological effects. In just 300 days, the effects are serious enough that Russian astronauts from the Mir space station suffered permanent bone and vascular-system damage and were barely able to walk upon landing. Psychologically, they were also a mess. To get to Mars requires a journey of over 500 days round trip, and probably a 700+ −day stay on the planet with its very low gravity. Psychologically, it is suspected that the profiles of the individuals who will do best living in cramped quarters for over two years are *not* those of any member of the current astronaut corps. One thing is certain: If we do attempt such a journey in the 21st century, we will make the voyage only when Mars is closest to Earth at "opposition." Also, we will make the trip at a time that minimizes the astronaut's risk for solar radiation and cosmic ray exposure. We know when the Mars oppositions will be, and we can estimate when the 21st-century solar cycles will be at their peak of activity. The earliest opportunity would be 2035, when the flare and cosmic-ray effects are about comparable but not at their peaks. To delay our landing beyond 2035, we may have to wait until 2065 as our last chance. After that, either the cosmic ray or flare conditions may prove unacceptable for making the attempt in the rest of this century.

15   When is the Sun predicted to go nova?

The Sun will never go nova. A nova explosion requires that the star have a nearby companion star that is a white dwarf, and that the primary star is an evolved red giant shedding mass. When enough of this mass is accreted by the white dwarf, it detonates on the white dwarf's surface to produce a nova (see Glossary). The Sun has no companion star close enough to make this happen. If it did, habitable planets like Earth would not exist. We also know that some stars produce super-flares, especially the very cool M-dwarfs. These flares could be luminous enough to melt the surfaces of the moons of Jupiter. The Sun does not appear to be the kind of star that would unexpectedly unleash such a lethal event. Lunar rock samples have tracks within their crystalline structure left over from energetic particles, and geologists have used this information to estimate that our Sun has not produced any flares that are ten times as bright as the biggest we have seen so far, over the last 100 million years.

16   When will the planets line up again?

If by this you mean in a straight line across the solar system as seen from the Sun, this won't happen for thousands of years. Other less extreme possibilities occur all the time. You can use the web-based orrery at (http://order. ph.utexas.edu/clock/) to search for these kinds of alignments among the inner planets. I found one involving the Sun, Mercury, Venus, Earth, and Mars on

September 11, 2003, that is pretty cool. There is another on November 7, 2006, that is even better. Some of the wackiest literature written by adults that you will ever read is spawned by nearly every major planetary conjunction. The May 5, 2000, End of the World Alignment is perhaps the most famous in recent history. In the months before this rather unspectacular conjunction, people were forecasting the tidal disruption of Earth, a Great Spiritual Renewal, bouts of strange weather, and major devastating earthquakes. None of these happened, of course. (See Question 150.)

17   Has Pluto been demoted to the status of a non-planet?

Not yet, and probably never. At worst, it will be given a dual status. Astronomers who study the far solar system say that if Pluto (see Figure 3) had been discovered today, it would not be called a planet but instead a Kuiper Belt Object (KBO), which is what astronomers now call an object beyond the orbit of Neptune (see Question 150 for details). Since 1992, when astronomers first began uncovering the tiny faint streaks of these distant chunks of ice and rock in deep photographs of the sky, Pluto (1,413 miles across) has been rivaled by Quaoar (780 miles across) and Varuna (550 miles across). As the number of objects has swelled to over 600, Pluto

FIGURE 3 Pluto and Charon as viewed by the Hubble Space Telescope from a distance of nearly three billion miles. (CREDIT: DR. R. ALBRECHT, ESA/ESO AND NASA)

still looks impressive, but in the future who knows what we will find? A few astronomers such as Brian Marsden at the International Minor Planets Center have made the change already and don't even mention Pluto as a planet anymore. Other astronomers, however, have considered lowering the bar for what we call a planet to objects big enough to be round in shape. If that definition is adopted, not only will Pluto be a planet but also at least four other objects will be added to the list. A solar system with an "unlucky" 13 planets, however, might get at least a few people worried. (See Question 132.)

18   Which is the nearest planetary system to the Sun?

In 1998, the nearest star around which a planetary candidate had been detected was Lalande 21185, an M2-class red dwarf (visual magnitude +7.5) at a distance of 7.5 light-years from the Sun. It was thought to have two planets with masses of 0.9 and 1.6 Jupiters, at distances of 2 and 10 astronomical units (AUs) (Lalande 21185b and Lalande 21185c). These planets were detected, not spectroscopically but by the slight wiggles they caused in the motion of their parent star. Because of limited data, and the fact that the discoverers have not as yet published their data, this system is classified as "unconfirmed" at the present time. In terms of a confirmed planetary system, Epsilon Eridani, with its tally of two detected planets (Epsilon Eridani b and c), seems the safest bet at the moment. This is a K2-class dwarf star located 10.8 light-years from the Sun. The two Jupiter-sized planets orbit in highly elliptical paths, so that their temperatures must go through extremes even by the standards of Jupiter and Pluto in our own system, whose average distances they mimic. This could also be a problem for any smaller planets interior to the orbit of Epsilon Eridani b, although it is possible that smaller planets near or within an Earth-like orbit could remain stable. The so-called habitable zone for this star is between 0.4 and 0.6 AU, so this is far enough from Epsilon Eridani b that the prospects could be promising for a stable, water-bearing planet or satellite. The probable age of the star is about one billion years, making it a much younger planetary system than our own. It is known to have an orbiting dust disk at a distance of about 60 AU from the star. This star was also the target for a SETI search by astronomer Frank Drake in 1960. Science fiction fans will recall that this was the home star for *Star Trek*'s Mr. Spock. The *Babylon-5* station orbits the third planet in this system, which is also inhabited by the mysterious and ancient entity known as Draal. How the Vulcans could have overlooked such a mysterious entity in their own solar system is a mystery to me. (See Question 167.)

19   How do stars shine?

As a star forms, the in-falling gas gives up its kinetic energy by colliding with its neighbors. This heats up the gas in the core of the forming star. Once the temperatures reach about 10 million degrees, thermonuclear fusion between hydrogen atoms (actually protons) can start. Believe it or not, a particular proton will wait millions of years before it experiences a single fusion event to become a deuterium nucleus. The "hydrogen-fusion reaction" releases enough heat energy (pressure) to keep the star from collapsing further, and as a byproduct, a lot of light is also produced. The light, in the form of powerful gamma rays, makes its way to the star's surface, heating the gas, and producing a brilliant surface. The Sun emits nearly 400 trillion trillion watts of power. To do this, the fusion reactions have to convert 15 billion tons of matter into pure energy every hour. It's been doing this for about 4.5 billion years, and has lost only 0.02 percent of its total mass. When a star gets to be as small as 80 times the mass of Jupiter, it can no longer sustain hydrogen fusion in its core, and objects at about 20 times the mass of Jupiter cannot even sustain deuterium fusion, either. Objects between 20 and 80 Jupiter masses are called brown dwarfs. (See Question 128.)

20   When will the next galaxy hit the Milky Way?

Galaxies are so big, and their separations are so small in the universe, that collisions between them are actually rather common, especially in the dense cores of clusters of galaxies where they buzz around like bees in a swarm. The Milky Way is currently making a meal of the Large and Small Magellanic Clouds and will probably finish digesting them in about 500 million years or so. There are also several dwarf galaxies such as the Sagittarius System that are waiting in the wings as the next meal. The biggest, and most spectacular, collision will be with the Andromeda Galaxy, which will visit us in about three billion years, give or take a few millennia! It is approaching the Milky Way at a speed of roughly 700,000 miles per hour. By the time they are 300,000 light-years apart, powerful gravitational effects will tear spiral arms to shreds, and destroy the beautiful pinwheel shapes these two galaxies have had since they were first formed. There are many examples of these intergalactic "train wrecks" to be found in space. Plate 1 shows a particularly lovely example. What do you think the sky would look like with another galaxy plunging toward your own?

21   What happens near the event horizon?

The region of space very close to a black hole can be a very messy environment. Mathematically it is a very simple region dominated by the black hole's outer boundary called the event horizon. Matter is trying to flow into the black hole at the equator and, through friction, is being heated to thousands and perhaps even millions of degrees. Magnetic fields dragged in by the flow get amplified and concentrated. They may eventually pop out of the gas disk like solar prominences and flares, releasing bursts of X ray, and perhaps even gamma-ray, energy. Clumps of clouds and asteroids orbit the black hole in seconds and are shredded by gravity, to produce flickering quasi-periodic bursts of X-ray light as they slide into the horizon zone and are finally lost from our universe. Does fusion happen just outside a black hole? Not very easily. To produce thermonuclear reactions you need temperatures in excess of one million degrees to cause protons to collide with deuterium and produce tritium. Deuterium fusion is the lowest energy fusion reaction we know about. To get to one million degrees, protons have to collide with deuterium nuclei at speeds of about 290,000 miles per hour. Near stellar-mass black holes, gravitational tidal forces are substantially higher, and it might be possible for some of the gas to reach these kinds of conditions, at least in a limited volume of space near the horizon, perhaps even in "solar" flares that pop out of the magnetized accreting matter. A signature of this would be a black hole emitting bursts of gamma rays.

22   What is on the other side of a black hole?

We don't know for sure other than what theory tells us. Inside all black holes there is a region of extreme density where the matter that fell into the black hole is crushed into a so-called singularity state. The mathematics tell us that this region has infinite space-time curvature and gravitational field strength. Between the event horizon and the singularity, the mathematics show that time and space may actually "reverse" their roles, although what this might mean to an astronaut is unknown. Some predictions suggest that a worm hole might form inside. A worm hole is a particular solution to Einstein's equation for gravity in which two parts of space-time may be joined together. Many science fiction authors like to use them to allow spacecraft to travel quickly from place to place in our universe. But all of these ideas are based on "pure math" descriptions of how they might work, and, as you know, nature is often much messier than any idealistic, mathematical rendering. There are no perfectly straight lines in the universe, and there are not likely to be worm holes either. Worm holes do not exist naturally and would only form from certain types of black holes. So far as we understand the physics of real systems, black holes must have within them the physical object that formed them. As you enter a black hole, the object that formed it is still in front of you, filling the spacetime with gravitational radiation. Worm holes do not get born this way because there is always star-stuff blocking the doorway! (See Question 277.)

23   If you stuck your hand into a black hole, what would happen?

You can't just park outside a black hole and do this experiment because there are no stable orbits possible for objects within an "arm's length" of the black hole's event horizon. Your only option to get close enough is to plot a course that lets you fall into the black hole. General relativity says that as you approach the event horizon, nothing unusual is going on at all. All you would be feeling are the tremendous tidal forces of the black hole, and these could be very substantial across a distance equal to your arm's length. In fact, for a black hole produced by a star, by the time you are 40 miles from its horizon, the difference in the gravitational pull between your chest and hand is a thousand Earth gravities. Your hand would be torn from your body. But by then that would be the least of your problems. For the gargantuan black holes that lurk in the cores of many galaxies, the tidal forces near their event horizons are so minor that you might not even realize you had reached the event horizon at all, provided of course that the black hole was not surrounded by a lethal accretion disk spewing out X rays and gamma rays. You would gently pass across the horizon with your body intact, but with a very dismal future awaiting you as you continue to fall into the singularity (see Glossary) located a few billion miles away. In either case because of relativity, distant observers would see your hand "wink out" followed moments later by the rest of you!

24   How fast does gravity travel?

Before Albert Einstein (1879–1955) developed general relativity in 1917, Isaac Newton's theory of gravity required that gravity and gravitational forces traveled instantaneously from place to place. The most elementary assumption behind Einstein's theory is that gravity must travel at exactly the speed of light. In 1975, astronomers Joseph Taylor and Hendrik van der Hulst discovered two pulsars (see Glossary) orbiting each other at very close range. From a decade of careful study, they discovered this system collapsing, and the speed of this collapse is within 1 percent of what you would predict if gravity waves were carrying off energy at the speed of light. What has been claimed as the first direct measurement of the speed of gravity was attempted by Ed Fomalont of the National Radio Astronomy Observatory and Sergei Kopeikin from the University of Missouri in Columbia. On September 8, 2002, the planet Jupiter passed in front of a powerful quasar J0842+1835, and from the gravitational distortions that Jupiter's field produced on the passing radio waves, Fomalont and Kopeikin found that the speed of gravity was about equal to the speed of light. Several world-renowned experts on general relativity disputed this result and suggested that the astronomers had merely measured the speed of light, not gravity. We will likely not hear the end of this until the calculations are actually published and the rest of the community can go over them, too. Meanwhile another astronomer, Tom Van Flandern in Washington, D.C., has argued for many years that the speed of gravity is probably infinite. Van Flandern claims that if gravity traveled at the speed of light, the force upon Earth would be caused by the gravitational forces of the other planets based on where these planets were in the past, not where they are now. By including these delays, all the predictions seemed to be significantly off, and worse still, planetary orbits become unstable. Physicists quickly discovered that his calculations did not include a relativistic effect that exactly canceled the problems that Van Flandern had found. In the next ten years, more tests will be attempted, but for now "Einstein Rules."

25   What is the substance of gravity?

No one knows for sure, but it's not because we don't have any good ideas. We do know one thing: Einstein's theory of general relativity seems to be very accurate. Within the details of this theory there is a built-in, and very specific, way of interpreting what gravity is. In Einstein's own words, "Space-time does not claim existence on its own, but only as a structural quality of the gravitational field." What he is saying, and what the entire theory of general relativity is based upon, is the incredible idea that what we call space and time are not things that exist independently of the gravitational field of the cosmos. The gravitational field actually "creates" these qualities of the physical world. Physicists have had a lot of experience working with fields in nature, and from this they have developed ways of thinking about gravity that also give us some insight into how this field works. Theoreticians believe it is at its heart a "quantum field" just like the fields associated with the other three fundamental forces: electromagnetism and the strong and weak nuclear forces. It shares many subtle mathematical similarities with these fields that would be difficult to understand by random chance alone. This quantum field description that we use for the other forces also means that gravity would be a shimmering cloud of invisible particles; you can call them gravitons. In order that they add together to create the kind of universe we see, the gravitons have to carry a "spin" of 2 quantum units. Like the particles that produce the other forces, gravitons would have no rest mass. They would travel as packets of space curvature, at the speed of light. If you change any one of these qualities, gravity as a force will absolutely not look like the force we see on Earth, the solar system, and the rest of the cosmos. But these particles don't resemble the grains of sand on a beach. They don't have hard edges or shapes but have to have a spread-out and indistinct character to them, although they can also have a very precise set of mathematical properties. No one is really sure exactly what kind of description works best to describe them, and it doesn't seem we will ever be able to study one of these particles to prove or disprove the many descriptions already offered.

26   Is there such a thing as anti-gravity?

Every time a toy magnet attracts a paper clip it is producing a force that is "anti" gravity. When you do this experiment, you are working with a force that is stronger than the gravity produced by an entire planet acting on the same paper clip. We can create electromagnetic forces that are literally trillions of times stronger than gravitational forces from the same material objects. Gravity is, after all, the weakest force in the universe. The universe also contains a genuinely mysterious force that is stronger than gravity on a cosmic scale that we call dark energy. It isn't a force produced by any of the other processes we have identified under laboratory conditions, and frankly we have no idea what really causes it. It does operate in exactly the opposite direction to gravity, causing matter to feel a powerful repulsive force, but it only seems to act at the cosmic scale across millions or even billions of light-years. It seems to have no measurable action at galactic, solar-system, or planetary scales. Whatever causes it, it is a different kind of force than gravity, and it is not mathematically an "anti-gravity" force in the same way that polarity in magnetism or electricity produces an opposite force between like-polarity bodies. Gravity itself is incapable of producing repulsion because it is a purely attractive force. We have yet to see if it remains attractive between bodies made of anti-matter, but Einstein's description of gravity predicts that this should make no difference to gravity. Yevgeny Podkletnov in Finland got a lot of attention in the 1990's for a device he built that claimed to cancel several percent of a body's weight, but the result, like the claim for cold fusion in the 1980's, could never be reproduced. Another recent entry into this arena in 1996 has been James Woodward at Cal State–Fullerton. What he does is to vibrate an object in just the right phase so that its inertia is partly canceled— at least that is what he claims. This technique has received a lot of attention but mostly in the non-scientific press. These techniques try to shield gravity, or to change the definition of inertia. This is not, however, the way to make an actual "anti-gravity" force. To do this in the modern arena of quantum field theory, we have to start at the bottom. We have to propose a new carrier for forces whose properties are "opposite" to the hypothetical graviton. Particles with a quantum spin of zero produce vacuum forces that are repulsive. These are the particles that are candidates for dark energy, which we have ruled out as a genuine anti-gravity force. The only other hypothetical particle that would produce a genuine repulsive gravity-like force is the gravitino. This particle has been predicted by what physicists call supersymmetry theory. No one has detected the gravitino yet, but it is a member of a whole new raft of particles that are being hunted for at billion-dollar laboratories across the world, which are also searching for traces of the spin-zero particles that may be related to dark energy. If a genuine anti-gravity force exists, it will be uncovered in this way, not by trying to fool Mother Nature.

27 How old is the universe?

Since 1998, astronomers have narrowed down our uncertainty for this important number so that a value from 13 to 15 billion years is the most likely age. Astronomers are accustomed to quoting numbers in terms of ranges, but it has always been rather awkward to explain why we can't seem to know this number with, say, the same accuracy that we know the mass of the Sun, or the distance from the Sun to the center of the Milky Way. This all changed on February 11, 2003, when the Wilkinson Microwave Anisotropy Probe (WMAP) reported the results from its first-year's measurements of the cosmic microwave background. This feeble light, hiding out in the microwave portion of the electromagnetic spectrum, is all that remains of the fireball that was the big bang. Through a detailed study of the temperature irregularities in this light, NASA astronomer Charles Bennett and his colleagues delivered, as they had promised, a high-precision answer: 13.7 billion years with an uncertainty of only 100 million years. The WMAP results are completely consistent with ages determined from the oldest stars since the actual formation of stars and galaxies is now known from the WMAP analysis to have started no earlier than 200 million years after the big bang, or 13.7 billion years ago. This signifies the end of the so-called Dark Ages. Our Milky Way was probably among the first galaxies that formed as the Dark Ages came to an end.

28   What is the universe made of?

The recent successes of NASA's WMAP mission were announced on February 11, 2003. Among the measurements they made with extremely high precision was the overall composition of the cosmos. Since 1990, this had been known with fair accuracy, but now there is absolutely no more "wiggle room" when astronomers talk about the universe. Think of the full gravitational mass of the cosmos as equal to exactly $1.00. The WMAP measurements now say that 73 cents of this is in something called dark energy, 23 cents is in an equally mysterious "something" which astronomers call dark matter, and finally, the portion of the universe which is in the form of stars, gas planets, white dwarfs, and black holes is worth 4 cents. The next round of investigations will concentrate on a proper, independent, accounting of exactly what these three individual components might be.

29   What is dark matter?

By measuring the speeds of stars within galaxies, and the speeds of galaxies within distant clusters of galaxies, astronomers have become convinced over the decades that there is a sub-luminous or even non-luminous "something" that contributes to the gravitational fields of these systems but which you cannot count up optically so that the speeds and the masses "balance" each other. This is "dark matter" or "missing mass," and some physicists have proposed it may not even be matter. The first kind of dark matter we know about is just faint stars, distant white dwarfs, neutron stars, and black holes. These are, or once were, all made of ordinary matter (protons and neutrons) and have already been tallied in the weighing of the matter content of the cosmos. Next come neutrinos, which are known to carry a little mass but not enough to amount to more than a few percent of the dark matter. Finally, we have to deal with the unknown kind of dark matter that dominates the cosmos completely. We have no clue what this stuff is; only that it is probably a completely new type of particle that possesses some amount of mass and is slow-moving compared to light. In the ancient universe, it was cold enough that it formed gravity wells that the hotter, ordinary matter (proteins, neutrons, and electrons) eventually fell into to start clumping to form galaxies. Physicists have not seen any particle that would fit this new material, but imagine several candidates called axions, gravitinos, or simply "neutralinos." Until we see some traces of this stuff in our ground-based accelerator labs at Fermilab or CERN, we are pretty dead in the water in explaining what it could be.

30   What is anti-matter?

It is very much like ordinary matter, but it has the important distinction that the particle has exactly the opposite charge. The anti-matter electron (the positron) has a positive charge but has the same mass and spin as the ordinary electron. The anti-matter proton has a negative charge, not a positive one. But an anti-neutron has a zero charge, so why does it have an anti-particle? Because the proton and neutron are made from matter quarks, and we can just as easily assemble them from quarks made from anti-matter. Here's how to do it. For neutrons and protons there are two types of quarks that compose them; up quarks (U) and down quarks (D). A proton is made from two up quarks and one down quark (UUD), and a neutron is made from two down quarks and one up quark (DDU). The anti-proton consists of two anti-U quarks and one anti-D quark, which give a net charge of -1. A similar swap for the anti-neutron gives two anti-D and one anti-U quarks, which have a net charge of zero. So, anti-matter is more than just the reversal of the charge for a particle. If the particle is composite, like the proton and neutron, you have to build them from anti-quarks. Matter and anti-matter don't obey exactly the same laws of physics. The K (and B) mesons, for example, seem to decay in slightly different ways between the two types of matter. The decays tend to favor about 0.2 percent more matter than anti-matter in the end products. Cosmologists are very excited by this old discovery because it helps to explain why our universe is totally dominated by matter. Meanwhile, new experiments are testing the idea that matter and anti-matter are accelerated by gravity at exactly the same rates—a key prediction by general relativity. It's a tremendously difficult experiment to carry out, but if any differences are uncovered, it will be a problem for general relativity and all the ideas that have flowed from it in the last 90 years.

31   Is there such a thing as the multiverse?

I think so. And wouldn't that be an amazing perspective on our lives? The idea of a multiverse is pretty easy to describe. We live in a universe that shares the same natural laws, matter, forces, and properties for its vacuum. It is, technically, a single object in spacetime. But what if it were connected to other universes through black holes or worm holes? These other universes could be very different from ours, but separated from our own spacetime by perhaps a gulf of "other-dimensions." In a recent theory of matter and fields called M-theory, our spacetime is a single "3-brane" (see Glossary under "brane" and Question 306) that is embedded in a ten-dimensional vacuum just like a two-dimensional sheet of writing paper is embedded in the three-dimensional space of your room. Even if our braneworld is infinite, there are still many other possible braneworlds that could coexist in this 11-dimensional vacuum, communicating only by weak gravitational forces. Each of these braneworlds would be a separate universe, and the collection rightly termed a multiverse. One of the developers of inflationary cosmology, André Linde, considered this kind of scenario in the early 1980's when he considered that the structure of spacetime is a vast arena in which both successful and "failed" universes continue to coexist in a timeless state. They can have very different laws and physical circumstances interior to themselves—some permitting life, others stillborn. In 1997, astronomer Martin Rees in his popular book *Before the Beginning* seems to claim that he came up with this "new concept" of multiverse, but in fact this is not correct. In fact, a non-scientist came up with the term. Just as science fiction writer Arthur C. Clarke is credited with inventing geosynchronous satellites, the actual term multiverse can be credited to the work of another science fiction writer, Michael Moorcock, in his book *The Sundered Worlds* published by Paperback Library in 1962. Moorcock not only coined this term but also defined it in his story in a way that is identical to many of the theoretical definitions in vogue today. Here's what one of his characters, Renark, has to say about it in the story: "Many who have probed the perimeter of space outside the galaxy have mentioned that they have sensed something else, something not in keeping with any recognized natural laws. Others have had the illusion of sensing suns and worlds within the galaxy where suns and worlds just can't be. This has given rise to the theory of the multiverse; the multi-dimensional universe containing dozens of different universes, separated from each other by unknown dimensions." In modern M-theory, our three-dimensional braneworld would be separated from a neighboring one across a distance less than a few millimeters, but along a dimension beyond the usual three. It would seem that science and science fiction are not so different after all. It is not easy for the non-scientist to tell the difference. It's even impossible in some situations. (See Question 244.)

32 What competitors to the big bang theory are being tested today?

There simply does not exist a rival theory of the universe that is as comprehensive and as mathematically and physically articulate as big bang cosmology against which to compare big bang cosmology. This is identical to what happened to all of the various "theories" we have proposed to explain why Earth has its seasons. Eventually, after detailed and exhaustive study, only one explanation won. Today, the greatest activity is in comparing one version of big bang cosmology against another in which Inflation occurred, or comparing a big bang cosmology with a cosmological constant added. If you count up all of the variants of big bang cosmology, there are literally dozens, each predicting a different scenario for the evolution of galaxies and structure in the universe, or whether the universe is destined to collapse or not. These are all still considered big bang cosmology, but they differ in the selection of certain free parameters in the theory, which can only be decided by direct observation. In the last 70 years, there have been many seriously considered alternate theories to describing the large-scale properties of the universe, and even the origin process itself.

So here is the scorecard as of today:

- In DeSitter Cosmology (ca. 1917), the density of matter in the universe is zero. This theory doesn't look like our universe at all.
- Einstein Static Cosmology (ca. 1917) failed to agree with modern observations because the universe is expanding with time.
- Lemaitre Cosmology (1924), Steady State Cosmology (1950), Cold Big Bang Cosmology (1965): None of these explain the cosmic background radiation and the universal abundances of helium and deuterium.
- Oscillatory Big Bang Cosmology (ca 1930): The universe does not seem to have enough matter to make it a "closed" universe destined to collapse in the future.
- Brans-Dicke Big Bang Cosmology (ca. 1955) would cause the evolution of the Sun and the Earth-Moon system to be severely modified.
- Hagedorn Big Bang Cosmology (ca. 1968) was refuted by the discovery that quarks exist.
- Alfven Anti-matter Cosmology (ca. 1960) didn't satisfy the observational evidence in many ways. It turns out that no large-scale background radiation has ever been detected that can be attributed to proton or electron annihilation.
- In Chronometric Cosmology (ca. 1970): The expansion rate of the universe does not change the way that this theory predicted.

- Plasma Cosmology (ca. 1970) cannot explain the origin of the cosmic background radiation, its isotropy and temperature, and the abundances of helium and deuterium.
- Old Inflationary, Big Bang Cosmology (ca. 1980) didn't account for the fact that the cosmic background radiation is not as lumpy as this model would predict.
- In Cold and Hot Dark Matter Cosmology (ca. 1980), dark matter is not sufficient, of itself, to account for the accelerated expansion of the universe.
- Big Bang Cosmology with added Neutrino Families (ca. 1980) ran aground of the fact that there is no experimental evidence that more than three types of neutrinos exist.
- Brane Cosmology in less than five dimensions (ca. 1999) doesn't work because the force of gravity would not be an inverse-square law.
- Modified Newtonian Dynamics (ca. 1990) is not based on relativity, and it offers no explanation for cosmic background radiation, or even the expansion rate of the universe.

None of these ideas satisfy all of the observational evidence as well as inflationary big bang cosmology does. Cosmologists see no need to take large steps backward in our progress just to open the field to less capable ideas that in most cases have already died from neglect or disinterest or been fatally challenged by available evidence.

33   Where did the space come from that the universe expands into?

The simplest answer is that the space between galaxies "dilates" and does not come from anywhere. Another way to think about it is to imagine the universe as a kind of watermelon. When it is sliced perpendicular to its long axis, we can marvel at how the cross sections of each slice begin from a small wafer, expand to a maximum area, and then diminish in size as we reach the other end. General relativity says that the universe is a four-dimensional object and that gravity can only be described in its full four-dimensional terms. This is hard for us to do because we keep wanting to think of things happening in space at separate times. We insist on slicing gravity's field one time step at a time. At each time step, the volume of space increases, but in some sense, this space, like the pulp of the watermelon, has always been there. In each slice, the space does not "come" from anywhere, it is just revealed to us a moment at a time like the pulp in the slices of the watermelon. Space seems to appear out of nothingness, but in its full context as an object existing in both space and time, space has always been present and isn't arriving from anywhere.

34   What is the universe expanding into?

This is a popular question, and it comes about because astronomers have identified the universe as being of an infinite "big bang" type with a definite origin to space and time at 13.7 billion years ago. Also, the circumstances of its birth are summarized by inflationary cosmology, which makes predictions about what the universe was doing when it was a trillionth of a trillionth of a trillionth (10^{-36}) of a second old. The basic idea is that space and time were "created" at the big bang at the earliest computable time, 10^{-43} seconds, by some as-yet mysterious process called quantum tunneling, quantum fluctuations, or a variety of other names. What came out of this was a vast arena of space but undergoing quantum fluctuations in gravitational curvature and energy probably throughout its entire plenum (see Glossary)—we may never know for certain. This plenum, which contained highly curved and changing space curvature, expanded in time (a quality that now had meaning) until the universe got to an age of about 10^{-35} seconds. Then one of the quantum patches in this vast dilating and stretching sea of fluctuating curvature underwent inflation. It expanded from 10^{-24} inches to several inches across, yet it was still connected to the even vaster plenum that had emerged from the big bang. We don't know what this vast plenum was doing at that time. Some other parts of it might also have reached the inflation stage and dilated, too, but perhaps for shorter or longer periods. This plenum could literally have been infinite in the mathematical sense, or simply "very very big." Today, the universe continues to expand, and it is still connected to this vaster plenum "way out there" that emerged from the big bang, which is stupendously huge but whose limits we have no real way of determining. Our part of the universe is that region that inflated way back at 10^{-35} seconds, and today, compared to the size of the visible universe (13 billion light years in radius), this larger region could be 10^{50} or 10^{100} times bigger. The universe is expanding because the underlying, even vaster plenum is still expanding. Its limits are beyond our visible universe, and beyond even that huge patch that inflated, and encompasses all those other patches, which perhaps inflated, too, to make slightly different universes "way out there." But, even this larger universe isn't expanding by "conquering" or displacing already existing space. In a sense it is expanding by literally creating space as it goes. Beyond its limits could be pure nothingness lacking space, time, or dimension.

35   What existed before the big bang?

This is one of the Big Questions in modern cosmology. I think it is fair to say that we don't know for certain. What is really troubling is that we may never really know because we can never re-create these initial conditions. All we can do is launch educated guesses and conjectures based on where our best, or at least most interesting, theories seem to be leading us. We know that space and matter are two very different things, but long ago they were actually one and the same. Because Einstein's general theory of relativity works so well, we have to accept for now its description of space and time as only aspects of the gravitational field of everything in the universe . . . gas . . . energy . . . matter . . . light. Near the big bang, gravity amplified itself by feeding off of its own energy in a complex, contorted, and brief state that ended when this gravitational energy produced the first generations of particles and anti-particles. So, matter in its most elementary field-like state was once part of space. Space remains indistinct from gravity, and so everything we see around us was once part of the invisible field we call gravity, which flashed into existence billions of years ago. Like a car rolling downhill, the momentum of this event is still with us and drives the expansion of the universe, and the clumping of matter into galaxies, stars, planets, and ourselves!

36   Where does space end?

Mathematical infinity and physical infinity can be two different things. If the universe is infinite, which is what we now conclude from our observations, there is no edge to space. Astronomers only receive information from a small part of the universe we are actually living inside because light has only had 13.7 billion years since the big bang to do much traveling across the vast space of the present-day universe. We call this the "visible universe" just to keep in mind that the universe is actually much larger than the part we will ever be able to see. Beyond the visible universe is more of the gravitational field of the cosmos (and embedded galaxies) that also came into existence in our big bang. Inflationary cosmology says that if we travel far enough in this larger space, we will come to some kind of fuzzy boundary where our physical laws may change somewhat. This represents the current limits to the small patch of space that our cosmos emerged from before it was inflated to monstrous dimensions. (See Question 257.)

CHAPTER 2

The Earth and Moon—Our Cosmic Home

Nowhere in the universe will we ever find such a beautiful pair of worlds with one planet to nurture us, and a lonely satellite to light our ways through a forest on a dark night. As we enter the 21st century after a breathtaking century of discovery, it's sometimes hard to believe we have any questions left to ask about our own twin-planet system. Only a scientist can easily see that many of these questions are far from simple. Through subtleties of thought, and new treasure chests of data, each question is the axis of its own universe of exploration and wonderment. How old is our Earth? How will it change in the uncertain future? Why did life find this world so inviting? What will happen if . . . ? Our Moon is a part of our familiar celestial backyard. It has even known the footsteps of humanity on its asphalt-dark surface. We have driven our awkward-looking cars across its hills, searching for ancient rocks. We have known, firsthand, the human dimensions of its craters. And despite the gloomy fiscal outlook, we have to hope that we will return to it someday and again watch Earth rise in its starry skies.

◗ ● ◗

37 Has the tilt of Earth's rotation axis, 23.5 degrees, ever changed?

Everyone eventually learns that the rotation axis of Earth is tilted 23.5 degrees to its orbit. For generations, teachers and scientists have dutifully noted this number. First calculated in the Western world by Eudoxus (400–347 B.C.), its decimal value has remained unchanged in our textbooks. But, in nature, nothing ever stays the same for long, especially in the complex world of interplanetary gravitational tugs. On May 5, 1900, the tilt was 23.4515 degrees. On May 1, 2003, the tilt was 23.4235 degrees. Calculations of the gravitational influences of the Sun and Moon show that the tilt actually varies from 22.1 to 24.5 degrees over 41,000 years' time. On even longer time scales measured in millions of years, the changing pattern of continents on Earth's surface might cause even greater variation. In 1998, geologist George Williams at the University of Adelaide tried to account for various lines of geological and glacial evidence by proposing that about 600 million years ago, Earth was locked in a planet-girding ice age where glaciers and ice caps stretched all the way to the equator, nearly making life extinct. He called it "snowball earth." The movement of the continents had also set the stage for an imbalance in the surface masses on our spinning planet. Is it possible that our poles could have been tilted as much as 54 degrees? This controversial proposal is still being looked into today, but it helps to explain how you can have glaciers in the current tropical regions and ice-free areas near the poles.

38 Has El Niño really slowed down the rotation of Earth?

Yes it has. At the spring meeting of the American Geophysical Union in Boston on May 27–29, 1998, John Gipson of the NASA Goddard Space Flight Center showed that the climate effect known as El Niño had increased the length of the day by 600–700 microseconds (0.0006 to 0.0007 seconds). The length of day or LOD is monitored by using trans-Atlantic Very Long Base Line Interferometry, a technique pioneered and still used by astronomers to study distant quasars. During the 1982–1983 El Niño, an even stronger LOD increase of 800 microseconds was detected, with a milder LOD increase detected a year later of about 400 microseconds. To understand how air currents can affect Earth's rotation, you have to consider an ice skater on the ice doing a spin. If she changes how far out she holds her hands by just a little, the rapidity of her spin is affected. Air masses change their location on Earth, and their distance from Earth's center by a few miles, and they also carry millions of tons of air in clouds. It is easy to understand from this how, with conservation of angular momentum, the length of day is constantly changing at a level of a millisecond or so.

39 How and why does Earth spin?

Take the ice skater analogy: To execute a spin she has to launch herself into it somehow. For Earth, this means that either the matter that formed it was originally moving, or that some large body hit it while it was forming, and that incident imparted the whack to set it spinning. Both of these processes probably happened, because we can still see them going on in the solar system today, but vastly slowed down. We are pretty sure that Earth was bombarded by large asteroids and planetoids hundreds of miles in size because we see cratering from such bodies all across the solar system. We think that the last major impact with an object nearly as big as Mars not only tipped Earth to its present inclination but also tore material out of our planet and formed a ring around it. A part of this ring condensed to become the Moon within the first 100 million years of Earth's formation. The impact energy liquefied the entire surface of Earth perhaps 4.4 billion years ago. It is amazing that the first fossils are present in rocks 3.8 billion years ago. Life was hell-bent on forming on this planet so soon after our planet had suffered a catastrophic impact.

40 What is the speed of Earth's rotation?

Because Earth is a solid body, not a liquid or gaseous one, all parts of Earth have to move around the axis a full 360 degrees every day. But, the speed of the planet at its surface depends on how many miles the surface moves in space to complete a 360-degree rotation in 24 hours. Near the equator, it has to travel its full circumference at that latitude, which is $2\pi R$, where R is Earth's radius. However, at any other latitude, Earth's circumference decreases by the cosine of the latitude so that you get $2\pi R[\text{cosine(latitude)}]$. You can see this with a piece of string and a globe. At the equator, Earth's circumference is 25,600 miles, and the day is 24 hours long so the speed of rotation is 1,070 miles per hour. This decreases by the cosine of your latitude so that at a latitude of 45 degrees, cosine(45) = 0.707 and the speed is 0.707 × 1,070 = 755 miles per hour. You can use this formula to find the speed of rotation at any latitude.

41 Did Earth spin faster in the past?

Well, we can't compare the passage of daytime to any clock that is based on Earth's own movement, so we instead compare a day to the ticking of atomic clocks, which have nothing to do with the Sun, Moon, and stars. When we do this, it is easy to see that the daily increase of the length of day is about 0.0015 seconds each century. That's why you sometimes hear about a second being added to the calendar on New Year's Eve every few years. Figure 4 shows how this slowdown has affected us since 1964. Half of this slowdown has to do with the tidal braking of the Moon. That's right, the Moon is literally putting the brakes on Earth's spinning. This also means that the Moon's

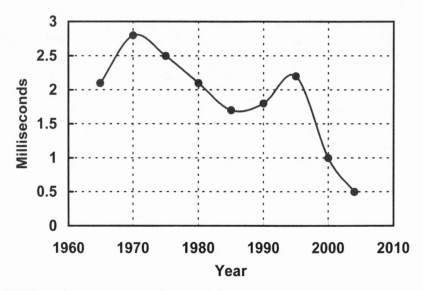

FIGURE 4 The days are certainly getting longer! Rotation slowdown of Earth since 1964 assembled from ultra-precise radio interferometry and other astronomical measurements, compared with Atomic Time. (CREDIT: INTERNATIONAL EARTH ROTATION SERVICES AND USNO)

orbit has to increase to conserve the momentum in the system, and this causes the Moon to drift farther and farther from Earth by an inch per year. As this process continues, it is predicted that in seven billion years, when the Sun becomes a red giant star, the lunar month will increase to about 47 days from its current 27.3 days. Earth's day will have lengthened from its present 24 hours to about 50 hours. Paleontologists have looked at fossils from 900 million years ago in the form of the growth rings in coral, seashells, and sediment deposits at the bottoms of lakes. These fossils all had layers called varves, which are related to the lunar cycle. The varve intervals indicated that the rotation period of Earth was about 18 hours back then.

42 Does Earth wobble as it spins?

Astronomers have been monitoring its movement for over 100 years and have found that Earth's axis dances in a roughly circular path called the Chandler Wobble. At the North Pole, if you were standing there and plotting where the axis emerges on the surface, its track would execute a circular loop about 20 feet in diameter during the course of about 433 days. This is about the pace of a snail with arthritis (one inch per week). This wobble is not the same year after year. The loop it makes under your feet doesn't reconnect with its own path like the orbit of a planet. In August 2000, Richard Gross, a geophysicist at the Jet Propulsion Laboratory, concluded the main reason that we have the Chandler Wobble is because of the changing pressure on the bottom of the ocean. These changes are caused by temperature, salinity, and variations in surface winds that alter ocean circulation patterns. Two-thirds of the Chandler Wobble is caused by these kinds of ocean-bottom pressure changes. The remaining one-third is caused by changes in atmospheric pressure.

43 Why doesn't the longest night of the year have the earliest sunset or the latest sunrise?

The Sun appears to travel faster in the sky when Earth is whipping around the Sun near its perihelion period in December. It moves more sluggishly during its aphelion period in June. Perihelion occurs around January 6th at a distance of 0.9832 AU. Earth's orbital speed is not constant during the December Solstice, and although it is about 1 degree per day, it varies by up to 0.3 degrees per day around this mean value. Because one degree of angle equals four minutes of time this translates into a rate of change of the sunrise and sunset times of up to a few minutes of time per day. There is also a major latitude effect because of the slant of the path taken by the Sun compared to the horizon.

44 Why is the latest sunrise in the year almost a week after the winter solstice?

If you were to observe the position of the Sun every day at exactly the same local time (not daylight savings time!) the Sun would trace out a figure-8 pattern called the analemma. The exact bottom of the analemma's winter loop marks the position of the Sun at the winter solstice when it arrives at a declination of +23.5 degrees. But the analemma only stands vertical to the horizon for an observer at Earth's equator. Only for observers along the equator will the latest sunrise (and the earliest sunset) always occur on exactly the winter solstice. For all other observers at other latitudes, the analemma is tilted on its side, so these times will be different.

45 Why are there more than 12 hours of daylight at the equinoxes?

You would think that if the Sun is located exactly on the celestial equator as it does on the dates of the equinoxes, that Earth would have exactly 12 hours of daylight and 12 hours of nighttime. The problem is that Earth is constantly moving in its orbit, which means that the Sun constantly moves in the sky each day. During the 12 hours of daylight, the Sun will move about 0.5 degrees, which equals about 2 minutes of time. This is enough so that when you try, then, to measure the length of the next 12 hours of nighttime, your number will not come out to be exactly 12 hours, because you are now 12 hours *past* the equinox!

46   What does Earth's core really look like?

For decades, textbooks have drawn pretty much the same picture of Earth's interior that showed it as a series of nested regions based on decades of seismographic data: the familiar outer crust lithosphere with its continents and oceans; the mantle where rock becomes plastic and flows in vast convective motions millions of years long. Further in we come to the core region. The outer core of liquid iron and nickel flows like some great electrical current of nearly a billion amperes creating our magnetic field. The inner core, discovered in 1936, is a solid iron-nickel body about 1,500 miles across. In 2000, two Columbia University scientists, Xiaodong Song and Paul Richards, announced that the inner core rotates slightly faster then Earth's surface. It's a very small difference, but in 400 years the inner core makes one complete rotation relative to the surface. Its spin axis is also about 10 degrees different from the spin axis of the rest of the planet, which could account for the offset between Earth's magnetic field and its rotation axis. In 2002, a team of scientists at Harvard University made an even more exciting, and controversial, discovery. Adam Dziewonski and graduate student Miaki Ishii studied 325,000 seismic waves that had passed directly through the center of Earth in the last 30 years. About 3,000 of these seismic events show a pronounced change as they penetrate even closer to the center. Their best model accounting for the velocity changes within the inner core is that there is a separate, 360-mile diameter, mini-planet in its center. They speculate that it may have been the original planetoid around which the rest of Earth accreted some 4.5 billion years ago.

47   Are there millions of mini-comets entering the atmosphere?

The current debate about this in the scientific community seems to be at an end. Since 1998, the scientific community has suspected that the dark spots seen in the ultraviolet images taken by the NASA Polar satellite were merely artifacts of the camera itself. Although it is true that a picture is worth a thousand words, for astronomers and physicists, convincing proof often has to clear other hurdles as well. No one can get over the sense of incredulity that these mini-comets can be so numerous, but leave no trace in the physical record other than a few dark pixels on satellite images. These mini-comets were calculated to evaporate hundreds of miles above Earth and to have sizes between a large van and a small house and densities below that of water. There would be about 20 of these entering the atmosphere every minute—10 million per year. This kind of inflow rate would be enough to give the Moon an atmosphere and cause seismic shocks easily detectable by the Apollo lunar seismometers. The plasma plumes in our atmosphere produced by these evaporating comets would have been detected by the military—and have not been. They would be visible to an observer of the night sky armed with a pair of binoculars. In order to avoid break-up at low altitude and flooding the stratosphere with their water, they would have to dissipate 600 to 15,000 miles above Earth, and nothing so concentrated as a 30-ton, 15-foot diameter "snowball" would actually break up at these altitudes, and certainly not from air friction. Finally, the rate of water deposition to Earth would eventually show up as an amount of water equal to about 10 percent the mass of Earth over the age of the solar system. This is enough to raise the oceans by 1/2 inch every decade and nothing like that has been observed. I think this idea is now formally a dead one in the scientific community.

48   What does Earth look like from far away?

I can't imagine a view more intriguing, or spiritual, than to see our Earth hanging like an indistinct point of light in space. The Apollo 8 astronauts gave us some spectacular photos of a blue Earth set against the blackness of space from the vantage point of the Moon. After a long hiatus, the history-making Voyager 1 spacecraft took the first "Earth-Moon" snapshot on September 18, 1977, raising the bar a bit on exotic photography. The following year, Galileo, on a gravity-assist fly-by of Earth en route to Jupiter, captured a respectable postcard photo. Clementine snapped its own image of "Earthrise over the Moon" in 1994 as a retrospective of the Apollo years. Our last intriguing image, shown in Plate 2, came by way of the NEAR spacecraft in 1998. The most curious and difficult tour-de-force came once again with Voyager in August 1995, which collected images of the planets from its vantage point beyond the orbit of Neptune. This montage, at times hard to read, shows mottled star fields and blemishes from data pushed to their limits, but the dim star-like images of six planets still show themselves. Earth is too faint to see, but its star field is counted just the same to show merely the sky in Earth's direction.

49 How many satellites orbit Earth?

Lots! And it depends on what you mean by a satellite. Usually we mean objects made by humans, fragments of metal, paint, tools, water, ice, and other odds and ends that are jettisoned or accidentally leave a shuttle or rocket. The Cheyenne Mountain Operations Center of the U.S. Space Command (http://www.cheyennemountain.af.mil/boxscore.htm) gives a box score of the total numbers of space vehicles and satellites it is currently tracking. As of June 11, 2003, there were a total of 9,126 objects of which 171 were satellites and 8,955 were classified as man-made debris including rocket bodies. Space Command has an ongoing program to track all artificial "satellites" larger than a few inches, and keeps an updated catalog of these objects along with predictions for their decay during each 60-day period. For example, STS-113, launched on November 24, 2002, is NORAD Catalog Object 27556. Although STS-113 eventually landed, it left behind various types of debris. The orbits of this debris constitute "artificial satellites" and, when added to all the other junk up there, pose a significant hazard to further space activities. By various accounts there are over 100,000 objects larger than 1/2 inch that have been tracked by NORAD in orbits that are mostly below 1,200 miles. Some of this material travels at speeds of up to 30,000 miles per hour. These include the debris from over 130 rocket and satellite explosions. In October 1999, the International Space Station had to be moved out of the way of a piece of orbiting junk that was headed right for it. Similar maneuvers have also taken place with the space shuttle.

50 Why is the sky blue?

The atmosphere is filled with minute dust particles. These particles are of about the same size as the wavelength of visible light, and they scatter light at the short blue wavelengths a bit better than they do the longer-wavelength red light. This means that as we look toward the sunset on the horizon, our path through the atmosphere intercepts lots of this dust, which preferentially scatters the blue component of the incoming sunlight out of our line of sight. The light is then reddened because the red light is not as easily scattered by the dust. As we look up toward the zenith over head, we are seeing not the sunlight coming directly from the Sun but the sunlight scattered by the dust at large angles to the line between the dust particle and the Sun. This light is blue because it contains little if any red light, which is not scattered as well by the dust. The rate at which blue replaces red light as you go from the horizon to the zenith at sunset and sunrise can be used to determine the optical properties of the dust and its size distribution.

51   How did life begin on Earth?

I am impressed by the fossil record that shows how single-celled life began 3.8 billion years ago, about 700 million years after the formation of Earth. Chemists have a ridiculously easy time of artificially creating pre-biotic organic molecules from the primordial atmosphere and water of the ancient Earth. In 2001, astronomers Lou Allamandola and Jason Dworkin at the Ames Research Center demonstrated that you can even create cell membranes and "vesicles" from common compounds and gases found in interstellar clouds. We know from geological evidence that free oxygen was absent in the atmosphere of Earth up until two billion years ago. We think that the atmosphere had abundant hydrogen-rich gases such as methane and ammonia because these are common gases in the atmospheres of the outer planets, and in interstellar space. Water has been detected in zircon crystals dating from four billion years ago. Laboratory experiments have shown that this combination of gases and liquid-water temperatures can be cooked into very complex organic molecules including amino acids, which are the building blocks for DNA. Large molecules resembling RNA may have formed on the clays that these organic-rich seas lapped up upon. After quintillions of trials on thousands of miles of continental shelf, self-replicating molecules formed and rapidly evolved in the food-rich oceans. The detailed steps may be hard to reproduce because Earth had millions of square miles and 700 million years to take the big step to bacterial life. Research labs have only a human lifetime, and a few square feet of test-tube surface! (See Question 319.)

52   Does the motion of Earth around its barycenter cause its core to slosh around?

The center of mass of the Earth-Moon system, called the barycenter, is the point about which Earth and Moon would seem to orbit if you were standing there. It is also this point that traces out the elliptical orbit of this binary system around the Sun every 365.24 days. Curiously, because the Moon is 81 times less massive than Earth, the Earth-Moon center of mass is not outside Earth's surface. It is located about 1,000 miles below its surface. Earth orbits this point every lunar month of 27 days, so the question is whether this motion causes any interesting geophysical phenomena. The answer seems to be "no," but there is an effect having to do with the lunar tidal force—the same one that causes ocean tides. There is a lunar tide that has an 18.6-year period and produces a 149-millisecond change on the length of day; a solar tide with a one-year period and 1.41 millisecond change, followed by a semiannual solar tide at six months with a 4.33 millisecond change. Following these primary tides are several minor tides including a lunar monthly tide with a period of 27.5 days and a change of 0.76 milliseconds. Apparently, the Moon does have a tidal effect upon Earth's rotation, but much of this is masked by the larger impact of the atmospheric variations so it's not easily to detect.

53 What is the maximum and minimum distance for Earth where the temperature conditions would be compatible with life?

We know that Venus, a twin to Earth in size, is thoroughly uninhabitable at a distance of 67 million miles from the Sun. Mars is a much smaller planet at a distance of 139 million miles, whose gravity is too small to hold onto much of an atmosphere. An Earth-sized planet at this distance, however, might still have livable surface conditions, especially if its atmosphere had some greenhouse gases like carbon dioxide. This would give it life-supporting temperatures over most of its surface. The temperature of Earth's surface with no greenhouse heating would be about -20 F. Water freezes at $+32$ F, so even at Earth's distance, greenhouse heating spells the difference between life and no life. If we were to take Earth, put it at the distance of Mars, with some greenhouse gases (mostly carbon dioxide), we could easily have liquid water at Mars's distance and even slightly beyond. So, I would estimate that Earth would become hospitable to life within a zone around a Sun-like star that extends from about 200 million miles to 80 million miles from the star.

54 Could an asteroid strike Earth with little warning?

We don't really know how many bodies are present in the range from 300 feet to a few miles in size. That's why astronomers such as the late Eugene Shoemaker (1928–1997) and his colleagues David Morrison and Donald Yeomans finally convinced NASA to support a systematic search for bodies in this size range. Every week they find more asteroids that are classified as Near-Earth Objects (NEOs) and as new data become available, the list of predicted closest approaches to Earth changes. Table 1 gives an example of a few of the asteroids with closest approaches to Earth during the 21st century. There are more than 500 NEOs known today, and the list is growing almost daily. What is most disturbing about the discovery process is that we only discover some of the NEOs after they have already made their closest approach. Typically, these objects are a few hundred yards to a half-mile across. It is these that could do local ecological damage, or kill a city. There are something like a few thousand bodies in this size range, and we only know of a few percent of these. They are very dim, and you only seem to spot them when they are less than a few days' travel from Earth because their angular speeds in the sky have to be high enough to see them move against all the other background clutter. We know the orbits of about 175,753 minor bodies in the solar system, but their orbits can and do get perturbed over decades and centuries. Two comets, Hyakutake and IRAS-Alcock, discovered in 1983 were nearly upon us before they were detected at all. Comets are a real problem, and can do nearly as much damage as asteroids. Hale-Bopp is one of the largest and is about 30 miles across. We knew about it many years before it dazzled our skies, because it was big enough to be tracked beyond the orbit of Mars. Had it been on collision course with Earth there is absolutely nothing we could have done about it. This is serious business, and the extent to which we all live on a knife-edge in relation to the devastation from these objects is very unsettling to those of us who keep track of these things. (See Question 8.)

55   Has anyone ever been hit by a meteor?

A study by Kevin Yau, Paul Weissman, and Donald Yeomans at NASA's Jet Propulsion Laboratory in 1993 turned up some surprising findings from old Chinese records going back 600 years. If you can trust these ancient chroniclers, there were four cases of human fatalities, and 13 cases of structural damage, during this time. Based on the current world population size, we should expect about one fatality every 60 years. Most recently, on March 27, 2003, a fireball streaked across the skies of southern Illinois and exploded near Park Forest. It produced hundreds of chunks of rock that damaged houses and even the insides of houses in the towns of Park Forest, Matteson, and Steger. According to a *Chicago Tribune* report,

> Another grayish, five-pound rock landed in the second-floor bedroom of Noe and Paulette Garza, of the 400 block of Indiana Avenue in Park Forest. "I could have gotten hit in the head by that thing," Garza, a 48-year-old

steelworker, said as he surveyed the damage caused when the rock crashed into his home about 12:30 A.M. The rock punched a hole through the roof and ceiling, shredded a set of venetian blinds, ricocheted off a metal window sill, shot about 15 feet across the bedroom and shattered a floor-to-ceiling mirror before coming to rest on the floor.

(See Table 12 for more examples.)

56 What would you do if you had one week's warning of an asteroid impact?

I would do something very primal, as would millions of other people. I would cry uncontrollably for a few hours, then I would pull my children out of school. I would thank God that I own a gas-frugal car and not an SUV, fill up my gas tank and drive as far north as I could into northern Canada. We would take back roads and listen to local road conditions on the radio to avoid fatal regions of traffic congestion. I would drive to stores along the way and stock up on basic supplies like bottled water, and I would bring my survival/camping gear with me. Our first stop would be in western Massachusetts to check up on my mother-in-law and family, then we would continue northwards with plenty of cash. My reasoning would be that it may be more likely for regions where the ecliptic plane rises high over head to be hit than places far from this zone because asteroid orbits tend to follow the ecliptic plane of the solar system. This means that locations within the temperate zone of Earth are likely to have a greater risk than arctic or sub-arctic regions. Of course, it is only an educated guess, and it might actually be better just to stay put where all our social services are located, and just cross our fingers, although being far away from inevitable urban crime waves seems like a good idea. For asteroids less than a mile across, only local regions are badly affected. It might make more sense to divide up the family and take off to two different geographic regions to maximize family survival. For larger asteroids, the effects are global no matter where you are. I would definitely not stay near the ocean where tidal waves would surely wash over us.

57 How are auroras produced?

For decades, physicists have understood the answer to this question, but textbook authors, especially of grade school science books, persist in giving the wrong answer. Auroras are not produced by solar particles that directly enter Earth's poles like water going down a drain. The actual reason is that it is something of a domino process where various phenomena have to happen in the right sequence. First a disturbance in the solar wind interacts with Earth's magnetic field. This causes a rearrangement event in Earth's magnetotail (the magnetotail is a region of Earth's field opposite the Sun that is stretched out like an invisible comet's tail). Then like a rubber band snapping, the magnetic field rearrangements in the magnetotail accelerate electrons, which are channeled to the north and south poles. A major sign of this disturbance in progress has been captured by the NASA Imager for Magnetosphere-to-Aurora Global Exploration (IMAGE) launched in 2000. Plate 3 shows that this satellite can see a flow of charged particles called the ring current grow in strength as the magnetosphere responds to the energy from the solar storm. Meanwhile, as the particles from the magnetotail pass into the ionosphere, they are boosted in energy to 6,000 volts. This million-ampere current slams into the atmosphere at elevations of 70 to 1,000 miles. The interaction between the electrons and the atoms causes the chief constituents, nitrogen and oxygen, to fluoresce like the gas inside fluorescent bulbs. The atoms emit light at only a few specific wavelengths, which give auroras their unique colors—blue, red, green.

58   If auroras didn't exist, how would Earth be affected?

Auroras are at the top of a complex pyramid of processes in nature. With them gone, some or all of these other processes would also be absent, with very broad impacts to Earth's space environment and the energy input to the upper atmosphere. Their absence would mean that Earth no longer had a magnetic field to protect itself from the full impact of charged particles from the Sun during solar flare events, and from the solar wind. The outer atmosphere would probably stop most of these charged particles, but the chemistry of the upper atmosphere and ozone layer would probably be affected. The ozone layer might disappear as other highly charged particles interacted with the fragile ozone molecules. Loss of ozone means more damaging ultraviolet radiation and increased cancer rates, but not the annihilation of the terrestrial biosphere.

59   During which months can auroras be most easily seen?

September and March are the most likely months for auroras, and January and July the least likely. This correlation was described by Jean-Jacques Mairan as long ago as 1733. Solar storms carry magnetic fields as they arrive near Earth. It is generally true that the more a solar wind's field is "south-directed" in polarity, the more vigorously it will transfer energy into Earth's magnetic field.

During the equinoxes, the Earth-Sun fields are oriented in a more opposed fashion than during the solstices, so this is thought to explain why auroras are so much more common and dramatic around September and March. It's just easier at that time for Earth's magnetic valve to open wider and let more solar plasma and energy into the magnetosphere. If you visit northern countries such as Alaska, Canada, and Norway during these times of the year, chances are good that you will see an aurora on practically any clear night.

60 How much mass does Earth gain from the solar wind and meteors?

Not much. The density of the solar wind is only a few dozen atoms per cubic inch traveling at a million miles an hour. This represents a flow of atoms into Earth of 180 tons a day. These particles are usually charged and would be deflected by Earth's magnetic field, so in fact very little of this gets into the atmosphere, perhaps only a few tons per day. Meteors and interplanetary dust add about 100 tons of matter to Earth every day and are a far bigger source of mass. Most of this is in microscopic dust gains (micrometeoroids), which you can actually collect in rainwater and study under a microscope.

61 What would Earth be like without a moon?

It would be pretty grim. In 1993, Jacques Laskar and his colleagues working at the Bureau des Longitudes in Paris performed extensive computer modeling of Earth's axis. What they turned up was that the Moon stabilizes the tilt of Earth's rotation axis so that we can have mild seasons with no extreme swings. Without the Moon, the gravitational torques from the Sun and other planets would cause the axis to wander and tip to extreme angles, even 90 degrees so that for a time it would roll on its side like Venus or Uranus. The fortuitous presence of the Moon so large and close has been argued by some scientists to indicate that the conditions for life elsewhere are rare. Without a big moon and the climate stability it brings, life can't evolve. Of course, we still have organisms that have adapted themselves to enormous extremes of temperatures, so perhaps life is more resilient than we think.

62 Are there any alternative theories about what causes the seasons?

This is a harder question to answer than you might suspect. The current explanation, having to do with Earth's axis tilt and changing sunlight angles through the year, has so thoroughly replaced any older ideas that you have to literally go back several thousand years before you begin to find traces of older ideas. Many of these have more to do with mythology than with any physical mechanism. The ancient cultures that created Stonehenge knew full well that the motion of the Sun in the sky was the deciding timepiece that went along with annual climate changes. The realization that it had to do with the tilt of the Earth's axis probably occurred around the time of the ancient Greeks. There are no alternative theories available today except as stories told by primitive societies. For us in our time, the explanation we use remains unchallenged for the simple reason that it is the correct one. There are also no alternative theories for why the Sun rises every day, why the sky is blue, how the Sun evolves in time, or even big bang cosmology! Some explanations are simply correct, and we should be thankful and not apologetic that humans have developed ways to find such correct explanations so efficiently.

63 What are the ten different ways that planet Earth can come to an end?

There are, of course, many ways that life could end over the course of millions or billions of years, but from moment to moment, there are very few plausible ways that nature would destroy Earth as a planet. In fact, there are none! We know where all the planet-sized objects are that, upon impact, could shatter Earth into an asteroid belt. These bodies are in orbits that are perfectly stable for the next billion years. The Sun will never go nova or supernova because it is the wrong kind of star to do this. It is also in a very stable part of its evolutionary phase, and studies of many other stars of the same type suggest only a slow, steady increase in luminosity over the next billion years or so. This will, however, cause Earth's oceans to start evaporating in about 500 million years—not an immediate crisis. There seem to be plenty of asteroids in our neck of the solar system with sizes of about one mile or so. If any one of these hit Earth, this wouldn't destroy it, but 65 million years ago, such an impact devastated the entire biosphere. A few billion people would die from starvation and other physical consequences. If this is the kind of "end" you mean, it is disturbingly likely in the next few million years. It could also happen in 100 years or less if we are unlucky. But it is unlikely that it will happen the day after you buy this book. (See Question 151.)

64 Are there differences in gravity from one place on Earth to another?

The local surface gravity at any spot on Earth depends on the density of rock at that location because, for a fixed volume of Earth, denser rock means locally more mass and so a stronger gravity. In regions where iron ore is abundant, the local gravity field would be 1 percent stronger than in regions with large deposits of sandstone or other comparatively lower-density silicates. As a satellite orbits Earth, its orbit can dip by several yards in response to the changing composition of Earth along the line from Earth's center to where the satellite is in its orbit at any moment. On the surface, a pendulum clock with a one-yard suspension would drift out of step with a more distant clock by about ten beats per hour.

65   How is Earth's distance to the Sun determined?

This distance, called the Astronomical Unit (AU), is the first rung of what astronomers call the fundamental distance ladder. It is this ladder of methods and phenomena that eventually lets us determine the distances to nearby stars, star clusters, galaxies, and quasars. Between 1639 and 1716, astronomers Jeremiah Horrocks (England), James Gregory (Scotland), and Edmund Halley (England) proposed that the transits of Venus could be used to precisely establish the size of the AU using simple triangulation techniques. This method was first used successfully during the 1761 transit, which yielded 95 million miles and was then refined during international observations of the 1874 and 1882 transits from which a refined distance of 92.7 million miles with an uncertainty of about 53,000 miles was eventually obtained. In 1931, a close passage to Earth of the asteroid 433 Eros led to a triangulated distance that greatly improved the value for the AU of 93 million miles. Alternately, you determine the orbit of an asteroid such as Eros and at a given date calculate its distance in units of the AU. Then you bounce a radar pulse off of the asteroid to get its distance in miles. You then know how many miles the distance of the asteroid in AUs corresponds to. The current adopted value for the AU is 149.598 million kilometers or 92,955,800 miles.

66   Where did all that nitrogen come from in Earth's atmosphere?

We think that the earliest "primordial" atmosphere had compounds that were similar to what astronomers have been finding in interstellar gas clouds, namely water, ammonia, methane, hydrogen, helium, argon, and cyanogen. This atmosphere was blasted away by the Sun's T-Tauri wind around 4.5 billion years ago, and we think a second atmosphere formed almost immediately from volcanic out-gassing. Titan, the satellite of Saturn, also has a dense, nitrogen-rich atmosphere. Water vapor, carbon dioxide, and nitrogen were the main constituents of the secondary atmosphere of Earth, as well as sulfur and other organic molecules, around 4.3 billion years ago. Reactions with the land and oceans eventually removed much of the carbon dioxide, leaving an atmosphere rich in nitrogen, which was soon joined by oxygen as bacteria began to respire oxygen.

67 What do we know about Earth's ancient magnetism?

The magnetism found in very old rocks indicates that Earth's magnetic field has been around for at least three billion years. However, based on the size and the electrical properties of Earth's core, if this field weren't being continually created, it would have vanished in only about 20,000 years or so. In addition, the polarity of the field has reversed many times in the past. The average time between reversals is roughly 300,000 years with individual reversal events taking only a few thousand years or less. What this all means is that there has to be a mechanism within Earth's interior that continually generates the field. It has long been speculated that this mechanism is a convective dynamo operating in our planet's fluid outer core, which surrounds its solid inner core, both being mainly composed of iron. The solid inner core is roughly the size of the Moon but at the temperature of the surface of the Sun. The convection in the fluid outer core is probably caused by buoyancy forces like the ones that cause fair-weather cumulus clouds to form on a hot summer day. These forces cause fluid to move outward from the center. The Coriolis "forces" due to Earth's rotation cause the fluid flows to be twisted into cyclones. Presumably this fluid motion twists and shears magnetic field lines, generating new magnetic fields. When Earth was younger and rotated faster, and there was more molten core material, and our field was several times stronger than it is today.

68 When will the next ice age arrive?

The last ice age ended about 12,000 years ago (10,000 B.C.). Since then, the moderate and warming global temperatures have allowed agriculture and human civilization to expand enormously. There have been episodes of cooling, such as the Maunder Minimum between 1645 and 1715 A.D., and the Sporier Minimum between 1460 and 1550 A.D., but these have been brief events in an otherwise mild climate period. 18,000 years ago the Wisconsonian Ice Sheet covered Canada and extended as a mile-thick glacier all the way to northern Pennsylvania, Ohio, Indiana, and Illinois. By 12,000 years ago the Bering Land Bridge had closed, ocean levels were on the rise, and Earth was growing warmer. By 11,600 years ago, a cold dry period called the Younger Dryas began a decline in plant and animal resources on which humans relied as hunter-gatherers. About 600 years later, ice core data show a steep rise in temperature. It appears that the bulk of the post–Ice Age warm-up may actually have happened in the space of just 40 years. The Pleistocene Era officially ended 10,300 years before the present (8,300 B.C.). Today we continue to live in what climatologists call the Holocene Interglacial Era warm period, but we know that ice ages do come in cycles with several different time scales. The great cycle measures 100,000 years and is driven by subtle changes in Earth's orbit, a discovery made by the Serbian astronomer Milutin Milankovitch (1879–1958). According to some recent research by Jan Hollan at the Nicolas Copernicus Observatory, forecasts of the Milankovitch Cycles show a continued warming trend. Orbital changes occur over thousands of years, and the climate system may also take thousands of years to respond to orbital changes. The models predict that the biggest stimulus of ice ages is the total summer radiation that is received in northern latitude zones near 65 degrees (65N), where major ice sheets have formed in the past. Astronomical calculations show that 65N summer insolation should increase gradually over the next 25,000 years. No 65N summer insolation declines sufficient to cause an ice age are expected in the next 100,000 years. In the short term, the forecasts of warming trends into the next few centuries imply continued increase in global temperatures, with a further reduction in mountain glacier cover. Already, many Canadian and Alaskan glaciers are in retreat. Dozens of glaciers present before 1950 have vanished. Alaskan glaciers have lost 500 cubic miles of ice and caused 10 percent of the eight-inch global rise in sea level since 1900. The snow cap on Mt. Kilimanjaro is expected to vanish permanently by 2015 due to global warming, which is not seen as part of the natural Holocene warming process.

69 What happens to a meteor when it hits the ground?

For small meteors, they reach a terminal speed of only a few hundred miles per hour, and by the time they reach ground they are cool to the touch. Large ones as big as a baseball will leave pits about as big as they are. Very large bodies a few yards across are not as easily decelerated by the atmosphere and may leave craters hundreds of yards across. They may be pulverized upon impact so that all that remains near, or in, the crater are fragments a few inches across. There may, or may not, be large chunks left over depending on the composition of the meteor. For the Barringer Crater in Arizona, only small fragments of a large body perhaps a hundred yards across are all that survives. The impact energy can be high enough to cause fires due to the frictional energy shed by the body. Most of the energy can propel lots of ground into the air, which rains down upon the surroundings in a thick blanket of heated, even molten, rock. Pieces of shaped glass called tektites are probably formed this way.

70   Has there been any new research on the Tunguska Event in 1908?

On June 30, 1908, an atmospheric explosion flattened trees near a region near Lake Baikal called Tunguska. Nomadic Tungus reindeer herders 20 miles away reported seeing a giant fireball and the destruction of forests. Interviews conducted several decades later described how thousands of reindeer were instantly vaporized. Forests within ten miles of the area were flattened as though a giant hand had descended from space and pushed them to the ground. The energy of this explosion has been estimated as over 2,000 times the Hiroshima hydrogen bomb. Since the first news of this event reached the Western world in the 1930's, a great many stories and theories have been offered to explain the event, and unfortunately scientists have often been the authors of the most outrageous ideas. People have suggested everything from a UFO engine explosion and anti-matter meteoroids to a rare magnetic monopole collision as a probable cause. Since 1999, new expeditions and computer simulations have severely narrowed the range of likely candidates. Estimates of how likely these events are have been revised from once every 200 years to once every 1,000 years. The best idea is that it was a loosely packed ice body from space perhaps 300 yards across that completely shattered into non-recoverable fragments some two miles above ground, creating a spectacular fireball and concussion wave. Modern acoustic "infrasound" monitoring stations have detected bodies up to 50 yards across entering our atmosphere, but they never make it to the ground. The fact that Tunguska wasn't observed as a fireball or bolide streaking across the sky outside this region implies an almost head-on collision with Earth over this region, rather than a typical "continent-girding" shallow trajectory. An international conference on Tunguska was convened in 2001 and included a wide variety of papers that entertained exotic explanations such as solar storm events and natural gas explosions as further possibilities that some feel are also consistent with the evidence from recent expeditions into this region. However, none of the original eyewitnesses are still living, and the region has substantially altered through natural processes over the last 90 years. Without any recovered material, this mystery may never be resolved to everyone's satisfaction—except mine. It's time to move on, folks!

71 What is the latest about the meteors that struck Greenland, Canada, and Texas?

On December 9, 1996, a very bright fireball streaked in a blinding flash over Southern Greenland. The magnitude of the fireball and, in particular, some seismic observations from Norway suggested that a large mass had fallen on the ice and possibly formed a crater. The seismic data were, however, later found not to be related to the fireball. Eyewitness reports and a security-camera video recording from Nuuk in Greenland made it possible to estimate an approximate position of a possible impact site above the glacier Frederikshåb Isblink, south of Nuuk. This area was searched many times beginning on January 4, 1998, using sophisticated synthetic aperture radar, but no traces were ever found of an impact nor were materials recovered from streams flowing out of the region. The Tycho Brahe expedition searched a large area above Frederikshåb Isblink in the summer of 1998 after most of the winter snow had melted. Unfortunately, no meteorites were found, and dust filtered from melt water did not include any material that could be related to the fireball.

On January 18, 2000, sensors aboard Department of Defense satellites detected the impact of a meteoroid near Whitehorse in the Yukon Territory, Canada. The object detonated at an altitude of 16 miles at 60.25 degrees north latitude, 134.65 degrees west longitude. The fireball was as bright as the Sun and streaked across the morning sky exploding with an estimated yield of 5–10,000 tons of TNT. The brilliant meteor attracted the attention of military satellites, seismic monitoring stations, and just about anyone who happened to be standing outdoors within 500 miles of the dazzling meteor's path. The object was probably about 20 feet across and weighed 200 to 250 tons. Hundreds of fragments were recovered of this "carbonaceous" material. Those pieces that had thawed turned into pools of dark organic sludge, so recovering the still-frozen-from-space pieces was a major discovery. This meteorite, one of the most ancient and primitive relics of our solar system, was later named the Tagash Lake Meteorite.

On Friday the 13th of October 2000, thousands of high school football spectators in Kansas, Oklahoma, and Texas were surprised when a brilliant meteor streaked overhead at approximately 7:30 P.M. Central Daylight Time. Sky watchers on the scene say that the slow-moving fireball was as bright as the full Moon. Satellite decay experts say the streak of light was probably a fragment from a Russian Proton rocket. Earlier on that day, a trio of Glonass satellites lifted off from the Baikonur Cosmodrome in Kazakstan aboard a single Proton rocket. Glonass is the Russian equivalent of the Global Positioning System. The successful launch added three new satellites to the Glonass array and dumped a fourth-stage casing from the Proton rocket over the skies of Kansas.

72 If Earth left orbit, would we tell immediately?

Yes, because the cause of this event would have been catastrophic. It would require either the disappearance of the Sun or a collision between Earth and some other planet-sized body. It could also happen if a star wandered close to the solar system and gravitationally upset the inner solar system balances with our Sun. If you missed this Big Event, you wouldn't be able to tell something had happened by looking at the nighttime stars with the naked eye because the parallax shift of Earth's position in space would be far too slight. But, astronomers would notice that the rising and setting times of the planets and Sun were becoming increasingly different every day. Computers that track the planets from observatories using telescopes would start to show tracking errors at the level of arc seconds or arc minutes depending on Earth's speed.

73 Is there an Earth-based telescope that can see the things left behind by the Apollo astronauts on the Moon?

The Apollo astronauts left behind a laser retro-reflector panel, and a ground-based telescope in Hawaii used a laser pulse to detect the reflected light from this panel in the early 1970's. It is still being used today to determine the precise distance to the Moon to better than 1/2-inch accuracy. As far as actually seeing anything through a telescope is concerned, it is impossible to do so from the ground. The lunar landing module launch pad was about 15 feet in size, which from Earth subtends an angle of 0.0027 arc seconds. The Hubble Space Telescope has a resolution limit of 0.047 arc seconds. This is the size of a dime seen at a distance of 10 miles or so. So, we are out of luck in seeing any traces of humans or human technology on the lunar surface—unless we look at the Apollo astronaut "home movies."

74   How far away from Earth was the Moon first formed?

Our best guess now seems to be that Earth was struck by an interloper about the size of Mars or so, which ejected material from Earth's crust into orbit forming a disk of gas. The Moon formed from this disk at a distance considerably less than its current distance by perhaps a factor of two. Since then, it has been slowly increasing its distance. The Moon may have been about 100,00 to 150,000 miles from Earth at the outset. By the time the Sun becomes a red giant, the Moon will be twice as far from Earth and half as big as it appears to us in today's sky. Meanwhile, the monstrously bloated Sun will fill up half the sky and bathe Earth in a furnace of 6,000 F.

75   Why don't all new moons cast a shadow on Earth?

The reason is that the shadow misses Earth completely. There are only two locations in the inclined lunar orbit where it crosses the ecliptic plane. At these points, the new moon shadow will fall on Earth and you get a solar eclipse. The rest of the time, the Moon is not in the right place at the right time, and the shadow misses Earth.

76   Why is the lunar month equal to the female menstrual cycle?

Well, it's not. Not really. Although the lunar "synodic month" from full moon to full moon is 29.53 days, the human female menstrual cycle can be anything from 26 days to 30 days, and some women are quite irregular. Chimpanzees average 37 days, humans 26–30 days, opossums 28 days, Macaque monkeys 24–26 days, cows 21 day, sows 20 days, sheep 16 days, guinea pigs 11 days, and rats 5 days. There is some tendency for large mammals to have longer menstrual cycles than small ones, and for primates to have longer cycles than other mammals. There is a very weak tendency for the cycle length to be correlated with mass, except for cows! Would a species of 1,000-pound primates have a cycle length of 30–35 days? We'll have to create one to find out.

77   Why does the Moon seem bigger on the horizon?

The Moon does not physically change its size, as any Apollo astronaut will be happy to tell you from first-hand experience. It also has nothing to do with the atmosphere, although this has been a popular explanation since the time of Aristotle. Instead, it has to do with how our brain uses information in the eye's visual field for finding clues about the sizes of things it is seeing.

When the Moon is near the horizon, there are trees, buildings, and other things that are used by the brain to give us an impression of how large the Moon is. But way up in the sky, there are no familiar objects for comparison, so the brain gives up and doesn't provide us with such a sense of proportion. It may also have to do with our eye's orientation in space as deduced by what the inner ear is telling the brain. There are many other visual illusions that our eye/brain system fall victim to. This particular one, called by the way, the Moon Illusion, is just one of many, and we are not sure we understand it fully. It also works for the Pleiades star cluster!

78 When will there be another full moon on Halloween or Christmas?

The full moon can be spectacular when it rises in the sky, especially on a significant calendar date. One of these dates is, of course, Halloween. A full moon on Halloween heightens its mystery and makes it easier for little ghouls and hobgoblins to safely find their way to your door. The next one will be on October 31, 2020. Although there is no religious significance to it, there will also be a full moon near Christmas Eve in the years 2004, 2007, 2015, 2026, 2034, 2042, and 2045. If you enjoy hunting for full moons around other holidays, try visiting the lunar phase calculator at the Data Services link of the U.S. Naval Observatory (http://aa.usno.navy.mil).

79 How hot is the Moon?

This is going to surprise you, but believe it or not the daytime temperature with the Sun overhead is about 265 Fahrenheit (130 C). Without the Sun, the nighttime temperature plunges to −170 Fahrenheit (−110 C). I am actually surprised by the large variations in the answer to this question that you will find on the web. Other estimates you will find: +107 C to −153 C, +127 C to −173 C, −200 F to +200 F, −193 C to 110 C, 400 K to 100 K. You would think that walking around on the lunar surface with temperatures 53 F above the boiling point of water would be a problem. Some "hoaxers" claim that there is no way that astronauts in space suits could ever survive in that heat, but they forget that space-walking shuttle astronauts experience exactly the same temperature extremes. During all of the Apollo missions, astronauts used space suits that allowed them to avoid being boiled alive. They had undergarments with a liquid cooling system that would move heat from the sunny side of the space suit to the colder shadowed area to keep them very comfortable. A malfunction in this cooling system, carried in their backpacks, would have been lethal.

80 How is it possible to see the Moon in the daytime?

There is nothing especially mysterious about this. It's just a matter of which is brighter, the Moon or the hazy atmosphere. As less and less of the Moon is illuminated, eventually you will reach a threshold where the sky is always brighter than the light reflected from the Moon, and the Moon will become invisible in the daytime. So long as the sky is clear of haze or high thin clouds, you can usually see the Moon and the Sun together in the sky at the same time if the Moon is in the right lunar phases.

81 Does the full moon have any effect on humans or animals?

None, except that, unlike animals, humans can talk themselves into making such correlations happen as self-fulfilling prophecies, or by selectively re-membering successes. Both of my daughters were born on the full moon, but the hospital nurseries were practically empty and no nurse made any com-ment about the full moon to me. Had the nursery been full, they would have made such comments and remembered them, thereby passing on the faulty observation to others. Ivan Kelly, a psychologist at the University of Saskatchewan, reviewed more than 50 published studies, which were based on over 200 different samples of data. For every positive study that claimed a correlation, there was a negative study that showed nothing going on. For ex-ample, two separate studies published in the December 23, 2000, issue of the British Medical Journal contradict each other on the question of whether animals bite people more during a full moon. Similar contradictions can be found when comparing many other studies as well, such as whether or not violence, police arrests, or self-poisoning cases actually increase during a full moon, according to Eric Chudler, a psychologist at the University of Wash-ington. The main issue is that people only remember when a full moon was present, and not all the other times when it was not. Also, it is a classic error in reasoning seen even in grade school that says that correlations imply cause-and-effect. Scientific reputations can fall when this distinction is not made. For non-scientists there is rarely such a penalty.

82 How rapidly does dust accumulate on the surface of the Moon?

The amount during the last three billion years has been about 4.5 yards, or a rate of one inch every 20 million years. Even before the Apollo moon land-ings, the consensus was building in the scientific community that very little dust would be found on the Moon and astronauts would not have to worry about sinking into the soil up to their eyeballs. The Surveyor 1 landing on

June 2, 1966, conclusively proved that a random location on the Moon had a rather solid surface with minimal dust. The Apollo astronauts reported lunar surface dust thickness ranging from one to three inches. Creationists often cite 15 million tons per year as a dust flow rate. Accurate values were actually measured in the 1960's as 22,000 tons per year. This rate, when extrapolated back over 4.5 billion years (assuming a constant rate), leads to a dust layer thickness on the Moon very similar to what the Apollo astronauts encountered.

83 Have any new lunar craters been spotted in the last 100 years?

Incredibly, the answer may be yes, although there is still considerable controversy over every recent sighting. On November 15, 1953, amateur astronomer Leon Stuart was photographing the moon and saw a bright pinpoint flash in the vicinity of the day-night terminator near the crater Pallas, shown in Figure 5. This discovery remained a curiosity of amateur

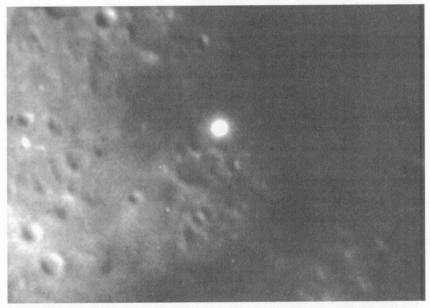

FIGURE 5 Lunar meteor impact caught on film. This dazzling, ten-second flash of light photographed in 1953 by amateur astronomer Leon Stuart has been confirmed by the Clementine satellite as a meteor that probably produced a young crater in this same region. Other astronomers claim it is a photograph of a "head-on" meteor—also a rare and exciting event seen during meteor showers. (CREDIT: LEON STUART. ENLARGEMENT BY LENNY ABBEY)

astronomers until data from the Clementine mission were carefully examined in 2002 by astronomers Bonnie Buratti at the Jet Propulsion Laboratory and Lane Johnson of Pomona College. They found a young impact crater just at the edge of visibility in the data. The crater is about one mile across. Clementine's resolution was about ten yards depending on orbit altitude. The crater required an explosive energy of about 500 kilotons of TNT to excavate, probably by a body about 60 feet across. Based on the size of the crater and impact-frequency tables, these kinds of events happen every 10 to 50 years or so on the Moon. Because there are far more amateur astronomers watching the Moon than professional astronomers, a lucky amateur astronomer will probably see the next impact, too. By the way, hundreds of flashes have been reported on the Moon since 1950. A ten-pound body striking the Moon should produce a visible flash, and it is estimated that as many as ten of these happen each year. Good times to watch for flashes are during meteor showers such as the Leonids.

84 Exactly how long does it take to get to the moon?

If you ask ten people to estimate an answer you might get times that range from a few hours to several weeks. Actually, it took the Apollo 8 astronauts 66 hours to get from the Moon to Earth. They set a speed record for this journey, which will not be bested until we return to the Moon. Apollo 10 took 76 hours from launch on Earth to lunar orbit insertion, including some hours spent in a parking orbit around Earth prior to trans-lunar orbit injection.

85 What dates will we be able to see the youngest crescent moon?

To view the youngest lunar crescent requires an unobstructed western horizon, clear air that is free of dust, and exceptional eye acuity for early twilight (eat plenty of carrots to pump up the rhodopsin level in your retina). The best times of the year occur when the path of the Moon in setting is essentially vertical. This usually happens in March for the northern hemisphere or September for the southern hemisphere. The record for seeing the crescent with a telescope was set by Jim Stamm on January 21, 1996, from Tucson, Arizona, when it was only 12 hours and 7 minutes old. With the naked eye, the record holder was Lizzie King from Scarborough, England, on May 2, 1916, when it was only 14 hours and 32 minutes old. These kinds of crescent moons are not even full crescents but merely a faint glow seen on the most western edge of the moon.

86   What is a pink moon?

The pink moon is the folklore name given to the full moon in the month of April. There are many names for each of the monthly full moons. A full moon in April is called the paschal moon or the flower moon in remembrance of growth and rejuvenation. May's can be called the frog return moon, or the merry moon (how about the merry frog moon?). The June "lover's moon" will light your way in the night and enchant and slide into the thunder moon by month's end. July's sturgeon moon lights your way through thickets of fireflies dancing in the grasses and hedges. The September singing moon will turn to the harvest moon if it strikes near the equinox. But by October it becomes the blood moon as the harvest and hunting for winter's larder commences. Once the cold sets in, November's frosty moon or mad moon turns into December's yule moon. The peacefulness of a winter's snowy scene may be illuminated by the silent moon, which becomes February's hunger moon if you didn't stock a proper larder, or a quickening moon if you feel up to watching the first subtle signs of an approaching spring. And finally we complete the cycle with the March's lenten, chaste, or seed moon as spring arrives and planting begins once again. There are also some names for the moon that occur in music and astrology. Although a lover's moon seems legitimate enough for June, bad moon is technically not the name of an identified phase, even though Credence Clearwater Revival made a big musical hit out of the name back in 1970. Added to this mix of recognized lunar phases are the astrologer's disseminating moon and balsamic moon, which occur just before the waning gibbous phase, and just after it. The dark moon, also called the new moon, is a time when the moon is "resting" between cycles and drawing up new ideas for the next cycle. Astrologers say you shouldn't undertake any critical activities when the moon is "void of course"—a time span of a few days when the moon is leaving one zodiacal sign and entering the next one. How ever you chose to name it, take the time to go outside some night and reacquaint yourself with the full moon's dazzling mystery and sublime beauty.

87   How long is the lunar month?

That actually depends on which points in its orbit, or in the sky, you use as a reference. The time it takes the Moon to return to the same spot in the sky relative to the stars is called the sidereal month and is 27.32166 days long. The time between, say, new moon and new moon is called the synodic month and is 29.530 days long. This is longer than the sidereal month because it is based on the geometry between the Sun, Earth, and Moon. During a lunar month, the Sun advances enough so that it takes another 2.2 days to catch up with the next repetition of the same phase. Then there is the nodal month, which is the time it takes the Moon to travel along its elliptical orbit from its ascending (or descending) node and return there, after 27.2122 days. There is also the anomalistic month of 27.554 days, which is how long it takes the Moon to travel from one perigee to the next. Finally there is the tropical month of 27.32158 days between one lunar equinox and the next lunar equinox. For all intents and purposes, the month is defined as either a 27.312 sidereal or a 29.530 synodic period.

88   Do lunar and solar eclipses have any noticeable effect on humans?

There is no evidence that eclipses have any physical effect on humans. However, eclipses have always been capable of producing profound psychological effects as every human who has ever seen one can tell you. For millennia, solar eclipses were considered to be portents of doom by virtually every known civilization. Although there are no direct physical effects involving known forces, the subjective "forces" brought on by the induced human psychological states have led to physical effects. These have stimulated responses that run the gamut from human sacrifices to feelings of awe and bewilderment to outright panic.

89   Why are some lunar eclipses red in color?

The color of a lunar eclipse depends entirely on the properties of Earth's atmosphere through which the Sun's light passes on its way to the Moon. The visibility of these colors on the ground also depends on local observing conditions. Sunlight passing through Earth's atmosphere loses some of its blue color as atmospheric dust scatters the blue light out of the light beam, leaving the red component mostly intact. This causes the light to take on a reddish cast.

90   How far does the Moon travel in its orbit?

The average radius of the Moon's orbit is 238,000 miles. The circumference is then 1.5 million miles as a circular path. The ellipticity of the actual orbit is so slight (about 5 percent) that the "circular" approximation is probably adequate for most purposes. Let's broaden the problem to include the planets.

Earth travels 580 million miles, Mars travels close to one billion miles, and distant Pluto travels 23 billion miles each orbit.

91 What is it about the Moon's surface composition that makes it such a brilliant object at full moon?

The lunar surface in the bright highland areas consists of anorthositic rock, which has been pounded to smithereens and reconstituted into what are called breccias. These are rocks that reveal smaller fragments fused together into larger conglomerates. On Earth, anorthosites are rare, but the Adirondack Mountains in North America are believed to be of this material. As for the brilliance of the full moon, anorthositic breccias have a grayish color and are not white. In fact, of the sunlight striking the lunar surface, only 7 percent is reflected, which means the Moon is actually a poor reflector of light. The reason it seems so bright at night is that your pupil is opened up quite wide making your eye far more sensitive to faint light levels at night than during daytime. This sensitivity makes up, in part, for the poor reflectivity of the Moon, so that it appears a bright object instead of a dull gray!

92 How long does a full moon last?

Technically it lasts only a few hours, during which time it is 100 percent fully illuminated, but to the naked eye, the Moon would appear essentially fully illuminated for the better part of a day.

CHAPTER 3

The Sun—
Our Day Star

Helios greets us each day with a canvas upon which to write a new chapter in our lives. A golden star; the incandescent engine of life on this world. Can there be any more pressing questions in our minds than "Where did it come from?" and "How long will it continue to nourish us?" These millennia-old questions are now shorn of their mystery in the hands of science. It seems a pity to couch the sun's existence in terms of mere matter and the clashing of mindless forces. Yet by relinquishing a bit of its poetry we now see it for the troubled spirit that it is. A simple storm of blazing gases rushes out from its disk, disabling satellites and intimidating us. Over the great course of eons, it swells to monstrous proportions and devours worlds. All this in the life of an average star, in its middle years, as it travels across the azure skies of our little planet.

◑ ● ◑

93   What are solar storms and how do they affect Earth?

The expression "solar storm" refers to a variety of eruptions of mass and energy from the surface of the Sun. Flares, prominences, sunspots, and coronal mass ejections (see Figure 6) are the common harbingers of solar activity. They all involve sudden releases of stored magnetic energy that accelerate the hot gases near the solar surface or corona. Sometimes these storms make it all the way to Earth or beyond by flowing along the Sun's magnetic field. When the charged material, called plasma, collides with Earth's magnetic field, it can cause particles and energy to flow, eventually finding their way into the upper atmosphere, causing the beautiful and mysterious auroras. These same plasmas can produce magnetic fields that modify Earth's own field and affect compass readings. The changing magnetic fields can also set into motion processes that cause electricity to flow in long pipelines or produce electrical surges in power grids leading to brownouts and blackouts. If you want to

FIGURE 6 A massive outpouring of plasma from the Sun. The black disk is the occulting disk used by the instrument to artificially eclipse the bright disk of the Sun. This "coronal mass ejection" event was observed on February 19, 2000, by the Solar Heliospheric Observatory (SOHO). When these clouds encounter Earth, intense auroras and charged particle "storms" materialize and can wreak havoc. (CREDIT: SOHO/LASCO NASA/ESA)

learn more about how solar storms can, and do, affect us, you might want to read my book *The 23rd Cycle: Learning to Live with a Stormy Star,* published in 2000 by Columbia University Press.

94 How can I find out if there is a solar storm going on right now?

There are many places on the Internet that give forecasts about today's space weather. My favorite is the NOAA Space Environment Center at http://www.sec.noaa.gov/SWN. This resource tells you about many different indicators of space weather storminess. You can also visit http://www.spaceweather.com, which also gives a summary of the solar and geophysical conditions, plus many other neat things like keeping track of asteroids and comets on close approach to Earth.

95 What has the current sunspot cycle been like so far?

The current solar activity cycle is Number 23, and it began in November 1996. It reached its first "peak" (solar maximum) around April 2000, and after a few months of decline, the sunspot activity became more active and reached a second peak around January 2002. It is currently declining to its low state (solar minimum). In terms of solar storm and flare activity, by March 23, 2003, there have been 67 solar flares of Class X1 or brighter. Between January 1996 and March 23, 2003, there have been 5,428 coronal mass ejections, of which 354 have been "halo-type." This means they were directed toward Earth. The Earth effects of this solar storminess have resulted in 12 magnetic storms with a severity of magnitude 8 or higher on a 9-point scale. There have been six Great Aurora episodes, where observers in the United States could see prominent auroral activity, and when observers elsewhere on the planet also saw similar activity. The most powerful aurora emissions observed by the NPOES satellites occurred on April 18, 2001, at a power level of 828 gigawatts. During the cycle, 84 auroral events were detected at power levels above 300 gigawatts. There have been at least eight satellites that failed due to solar-storm–suspected impacts, including Telstar 401, Adeos, and ASCA. Hundreds of less-fatal "satellite anomalies" have kept satellite operators on their toes. The total estimated cost of satellite outages is in excess of $2 billion, with some research satellites such as the Japanese ASCA costing $450 million alone. Of course, this doesn't include losses and problems with secret military satellites. There have been no electrical blackouts equivalent to the March 1989 Quebec outage, but it is believed that the major storm events on April 18, 2001, and July 15, 2000, could easily have triggered similar conditions had they had occurred under more favorable conditions. At the current rate of decline, which appears rather smooth, solar activity will reach sunspot minimum conditions around August 2006 for a cycle length of 10 years. We will then pick up activity with the start of Solar Cycle 24. As with the start of every sunspot cycle, past records show that solar cycles can end and begin abruptly, so there is always the chance that Cycle 24 will never start, and we would have conditions similar to the start of the previous Maunder Minimum, also referred to as the European "Little Ice Age."

96   Does sunspot activity have any effect on our weather?

If it has, the effect is so subtle that no human could ever detect it just by studying local weather conditions. Astronomer Sallie Baliunas at the Smithsonian Astrophysical Observatory noted in 1996 that if you plot the sunspot cycle during the last 100 years against the severity of northern hemisphere weather systems, you do see a very interesting, in-phase, correlation. But you have to look at weather over large areas of Earth, not just local precipitation patterns, to see the sunspot-weather connection. No one has a good idea what could be causing it.

97   Why do sunspots vary in an 11-year cycle?

Sunspots have been observed for thousands of years. Thanks to sophisticated telescopic studies we can study them in great detail, as the recent image in Plate 4 shows. Sunspots are several thousand degrees cooler than the solar surface, so they produce less light and appear dark. Samuel Heinrich Schwabe (1789–1875) discovered the sunspot cycle in 1843, and since then they have been studied carefully. Actually the cycle can vary from 6 to 13 years. When you consider that the solar magnetic field changes polarity at the end of each cycle, it takes two sunspot cycles (22 years) for a full magnetic field cycle to complete itself. As to why the sunspots follow a cyclical pattern in time, the speculation is that they are just the surface tracers of the so-called solar dynamo cycle, which regulates the entire magnetic field of the Sun. For reasons that we are just beginning to understand, magnetic dynamos such as the Sun's and Earth's have periodic or possibly quasi-periodic episodes of polarity change, perhaps followed by long periods where no cycles occur. Sunspots are regions where sub-surface magnetic fields are strong enough to pop up through the surface and billow out into stable magnetic storms. These last from days to months depending on what is going on down below.

98   Has the solar "missing neutrino" problem been solved?

We think so. Fusion reactions in the core of the Sun not only fuse hydrogen and deuterium into helium "ash" but also produce neutrinos, nearly massless particles that hardly interact with matter at all and travel near light speed. A calculation of the power of the Sun given its core temperature would predict a particular flow of neutrinos out of the core and into specially designed instruments here on Earth. For decades, physicists have monitored this flow and

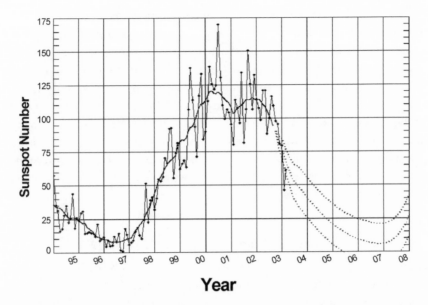

Year

FIGURE 7 Cycle 23 sunspot numbers through March 31, 2003. We are currently headed for a minimum in solar activity around 2006. Better forecasts will not be available until we are farther down from the peak activity period. (CREDIT: NOAA/SEC AND ISES)

consistently found that there are about half as many neutrinos coming from the Sun as expected. The implications for our understanding of the Sun and how it generates its energy are profound. Theoretical models predict that the neutrinos leaving the Sun depend on the actual temperature of the core, but the only way that these low numbers could arise would be if the core had, largely, stopped working at the expected high temperatures. In 2000, physicists from Canada, the United States, and the United Kingdom built another "heavy water" detector, the Sudbury Neutrino Observatory, over one mile below the surface in a nickel mine in Ontario. Unlike past experiments, the Sudbury detector is sensitive to not only the neutrinos generated by the nuclear fusion process, known as electron neutrinos, but two other types, called mu and tau neutrinos. Amazingly, in 2001, the Sudbury data showed that the total number of neutrinos leaving the Sun was equal to the rate predicted by theory, but what it was also saying very definitively was that some of the electron neutrinos were changing into the mu or tau neutrinos en route to Earth 8.5 minutes away.

Although the new results support our theoretical understanding of how the core of the Sun works, they challenge physicists in a profound way because these kinds of neutrino transformations are not allowed in the so-called standard model of the electromagnetic and weak interaction. The results also provide some insights into cosmology because the number of neutrinos expected to exist in the universe is determined by big bang theory. If you multiply the expected numbers by the possible masses of these "oscillating" neutrinos, you end up with neutrinos in the universe accounting for as much cosmological mass as all the visible stars in the universe. This is, however, far less than the dark matter component, so massive neutrinos still don't account for this even greater mystery in astronomy.

99 How do we know what the inside of the Sun is like?

Up until 1998, astronomers had virtually no way to probe the interior of the Sun, which was hidden behind a screen of plasma over 9,900 K hot at its surface. But the problem astronomers face is not much more difficult than what geologists are confronted with in exploring the interior of Earth. Thanks to the thousands of neutrinos detected by the SuperKamiokanda Neutrino Detector in Japan, and their trajectories, we can trace them back to the sky and "image" the Sun by the neutrinos streaming out from its core. An example of this neutrino image is shown in Plate 5. In time, we will improve this ability until we can focus closer and closer to the core itself. Is it a lumpy convecting furnace or completely round and smooth? Another method is to measure the slight speed changes in the solar surface as it adjusts to sound waves traveling through its interior. The surface information can be decoded into a picture of the solar interior, just as seismic waves from earthquakes help geologists dissect Earth's interior. During the last five years, the MIDI instrument onboard the SOHO satellite and the GONG program here on Earth have monitored the solar surface using instruments that show how the surface is changing from minute to minute. After many years, they have achieved the ability to define the basic elements of the solar interior, the thickness of its convection zone, the various jet streams that flow, and the boundary layers where velocity shears are formed. A by-product of this helioseismology technique is that we can now detect major solar storms on the backside of the Sun weeks before they rotate around to our side and pose a problem for us. Like gigantic earthquakes on the solar surface, these storms send sound waves through the solar interior, which affect the speeds of the gases on the side of the Sun facing us.

100 When can I see the next solar eclipse from North America?

Thanks to Fred Espenak at the NASA Goddard Space Flight Center, we can plan ahead for these events and where best to be at the time they are scheduled. His map is shown in Figure 8 for all solar eclipses in the 21st century. After a very long dry spell, on August 21, 2017, there will be one whose track goes from the state of Washington, passes down through Kentucky, and exits on the east coast near the Carolinas. There will be total solar eclipse through Texas, Kentucky, Ohio, and New England on April 8, 2024. The crossing of these two tracks in western Kentucky and southern Illinois will no doubt be considered "cosmic" for folks living in these regions. Further on into the 21st century, there will be total eclipses on August 12, 2045 (California, Colorado, Florida), March 30, 2052 (Georgia), May 11, 2078 (Louisiana, Alabama, North Carolina), May 1, 2079 (Pennsylvania, New York, Massachusetts, Maine), and September 14, 2099 (North Dakota, Michigan, Virginia). There are also eight annular solar eclipses, but those will hardly be noticed by the general public.

101 Will Earth ever fall into the Sun?

Earth is in an orbit that has been stable for over four billion years. There is no sign that it will not continue its predictable orbits for at least the next five

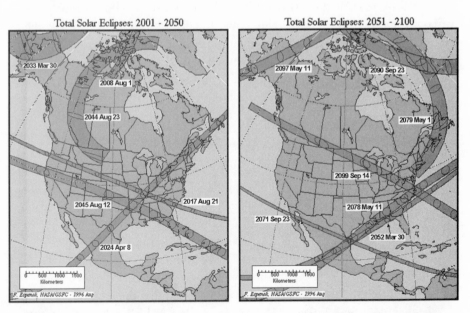

FIGURE 8 Total solar eclipses over North America during the 21st century. (CREDIT: FRED ESPENAK/GSFC)

billion years. By that time the Sun will begin to enter its red giant phase. Rather than Earth falling into the Sun, the Sun will expand out to meet Earth and probably begin to evaporate it into a puff of silicate and iron "smoke" by a billion years after that! A greater likelihood, however, is that Earth will be ejected from the Sun by a passing star. For four billion years we have been pretty lucky, even though there have probably been thousands of close passes by stars. In 1.5 million years the star Gleise 710 will pass within a light-year of the Sun. In the seven billion years remaining to our Sun, there will surely be many close encounters where the orbit of Pluto or perhaps of the inner planets is disturbed by the gravitational field of a passing star.

102 Could ancient astrologers have predicted total solar eclipses?

Ancient observations of solar eclipses from many different cultures and civilizations date back to at least 2500 B.C. in the writings that have survived from ancient China and Babylon. To establish an accurate calendar, people in ancient civilizations observed the moon regularly. Lunar eclipses were the first major celestial events that astrologers learned how to predict based on local historical observation records. At the same time that people kept track of how the lunar and solar calendars meshed with each other, they also uncovered some of the factors that lead to lunar and solar eclipses, which also require specific timings of the solar and lunar positions across the sky and over the years. By 2300 B.C., ancient Chinese astrologers already had sophisticated observatory buildings. Observing total solar eclipses was a major element of forecasting the future health and successes of the emperor, and astrologers were left with the onerous task of trying to anticipate when these events might occur. Failure to get the prediction right in at least one recorded case in 2300 B.C. resulted in the beheading of two astrologers. By about 20 B.C., surviving documents show that Chinese astrologers understood what caused eclipses, and by 8 B.C. some predictions of total solar eclipse were made using the 135-month recurrence period. By A.D. 206 Chinese astrologers could predict solar eclipses by analyzing the Moon's motion. The historian Herodotus (460 B.C.) mentions that Thales was able to predict the year when a total solar eclipse would occur. Details of how this prediction was made do not survive. The eclipse occurred in either 610 B.C. or 585 B.C. Apparently the method used worked only once because what is known of Greek scientific history does not suggest that the method was ever reliably used again. Ptolemy (150 A.D.) described in his book *Almagest* how he had a sophisticated scheme for predicting both lunar and solar eclipses.

Ptolemy knew, for example, the details of the orbit of the Moon including its nodal points. He also knew that the Sun must be within 20 degrees 41′ of the node point, and that up to two solar eclipses could occur within seven months in the same part of the world. Indian astronomy is largely wrapped up in the Vedic religious treatises, but one individual, Aryabhata of Kusumapura, born in A.D. 476, described numerical and geometric rules for eclipse calculations.

103   How did the Sun's brightness change during its first few billion years?

Although its brightness seems pretty constant today, it did change in a very complicated way while it was still forming from in-falling gas and dust. It was actually tens to hundreds of times more luminous, but it shone mostly as a powerful infrared "star" rather than one that could be seen by its optical light. Then when its nuclear fusion reactions kicked in and dust cleared, stellar evolution models show that it began to stabilize within a few million years as a star with a power about 70 percent of its current output. Soon after it formed, the Sun was actually dimmer than it is today. It has grown steadily in power as it has evolved as a star, and will continue to increase in power for the next seven billion years. During most of this time, the power will increase by about 1 percent every 300 million years. In the distant past, our atmosphere was loaded with greenhouse gases such as carbon dioxide and methane so these gases could have kept water in a liquid form on Earth's surface even when the Sun was 30 percent dimmer than today. By 500 million years from now, the power of the Sun will be several percent higher. This will cause extensive heating of Earth, the destruction of the biosphere, and by one billion years from now, the evaporation of the oceans into a Venus-like atmosphere.

104   How old is the Sun?

The Sun is estimated to be about 10 to 20 million years older than Earth, and by radioactive dating, Earth is about 4.5 billion years old, so the Sun is therefore around 4.51 billion years old. Note that some of the oldest meteorites are 4.8 to 4.9 billion years old, but the dust grains trapped in these meteorites can be much older than the Sun because dust grains formed in old stars tens to hundreds of millions of years before they got to the cloud out of which the meteorites, Suns, and planets formed.

105 Is the Sun shrinking in size by 200 miles a year?

No, and you can prove this with a little math. The solar radius is 441,000 miles, and it is at a distance of 93 million miles from Earth. In the sky, it appears to be a disk with an angular diameter of about 30 arc minutes, so 200 miles is an angular change per year of (30 minutes × 60 seconds/minute × 200/ 441,000) = 0.8 arc seconds per year. If this has been going on for the last 50 years, it would have changed the angular diameter of the Sun by 40 arc seconds or a little under 3 percent of its apparent size. This means that 50 years ago, there would have been no total solar eclipses. But we know people have been seeing them for thousands of years, so the Sun is not shrinking at a rate as large as this. In fact, the Sun's radius is slowly growing in size as it ages. This rate is unmeasurable. During the next five billion years, as the Sun continues to age, its radius will increase by one inch every two years. It will take 100,000 years for this natural rate to be measurable at all.

106 Is there any proof that the Sun is a member of a binary star system?

There have been many searches and proposals for a companion to our Sun. The most famous of these was in 1984 when astronomers David Raup and John Sepkoski (1948–1999) at the University of Chicago proposed that there was a 26-million-year period to extinctions on Earth. They went on to claim that a companion to our Sun called Nemesis was located 8 trillion miles away (88,000 AU) in the distant Oort Cloud, and that it might be sending a hailstorm of comets toward Earth every 26 million years or so. To avoid visual detection, the object would have to be a dwarf star or an inactive neutron star. It could not be a normal star. Extensive computer modeling later showed that a companion at this distance would be highly unstable and vulnerable to being ripped away by passing stars. In 2000, astronomer John Matese at the University of Louisiana rekindled some interest in this search by announcing that a statistical study of the orbits of long-period comets showed an unusual bunching effect. He interpreted this as evidence that something is "out there" perturbing the orbits in synchrony. The best estimate for a culprit would be a brown dwarf with a mass three times Jupiter's in an orbit with a radius of two trillion miles (25,000 AU). So far, nothing like this has been detected telescopically, although this object, sight unseen, has been adopted by many fringe groups who believe it is Nibiru, Wormwood, or other similar objects that portend the End of the World. (See Question 360.)

107　What does the Sun look like from the surface of Mars?

The wonderful images returned from the Pathfinder/Sojourner lander on Mars on August 2, 1999, showed humans for the first time what sunsets on this planet look like. Many versions of these images have appeared in press, especially the red "false color" ones, but these are not accurate at all. They correspond to our bias in thinking that the red planet should have a blood-red sunset. In fact, a true-color version of the Sojourner Mars sunset is also available that more accurately shows what sunset would look like to human eyes. The sky is black, changing to brownish gray as your eye scans toward the setting Sun. The sky near the Sun is a pale blue color, and the scattering of the sunlight in this near-airless environment laden with dust produces an unusual fan-like illumination perpendicular to the horizon and reaching toward the zenith, with the Sun at its vertex. The azimuth extent is 60°, and the elevation extent is approximately 12°. The lighting and cameras were not sensitive enough to see stars in the twilight sky of Mars, but they were most assuredly already present in the dark sky behind Sojourner and would have been detected if it had had a camera sensitive enough to see them.

108　How far is the Sun from the center of the Milky Way?

When I started out as an astronomer in the early 1980's, the adopted distance was 32,600 light-years, but even then there were plans to revise this number downward, and to measure it more precisely. This number, like the distance from the Sun to Earth, is a linchpin quantity in astronomy that establishes the scale of the universe. A major review of the various estimates for it was published in 1993 by astronomer Mark Reid at the Smithsonian Center for Astrophysics, who found that the best value was probably closer to 25,000 light-years give or take 1,600 light-years. The most recent value determined by Harold McNamara and his colleagues at Brigham Young University is 25,700 with an uncertainty of 980 light-years.

109　How often does the Sun pass through the spiral arm it was born in?

Our Milky Way looks like a giant pinwheel in space, but the arms that make up this spiral are not solid objects that rotate around the hub of the Milky Way. Instead, think of the "wave" you might be a part of at your local baseball stadium. Every few million years, as one of these arms advances in the

Milky Way, it involves a different collection of clouds, gas, and stars in the disk of the Milky Way. There are about two main arms that cross the solar circle, and since the Sun takes about 240 million years to orbit the center, this means that every 110 million years or so, the Sun enters a new arm system. But a lot has changed in the Milky Way in the 26 galactic years that have come and gone since the Sun was first born. We may never find ourselves in exactly the same neighborhood of our Sun's youth. Everyone has moved away.

110   How many stars are there like the Sun?

A lot!! If you just ask how many stars there are in our neighborhood of the Milky Way that have the same color (spectral class G) and luminosity (Main Sequence class V), there are 0.0063 of these stars per cubic parsec. This doesn't sound like a heck of a lot, but if you were to consider Earth sitting at the center of a sphere extending just 324 light-years in radius, the volume equals 4,188,000 cubic parsecs so that 4,188,000 times 0.0063 equals 26,384 stars similar to the Sun. In terms of an exact match, the Sun is classified by its temperature (9,900 F) and its power as a G2 V star. But any planet located at a distance of 93 million miles from such a star would see the same yellow star with the same amount of heating at its surface. The nearest G2 V star to the Sun is Alpha Centauri at a distance of 4.3 light years, which sadly doesn't seem to have any planets at all, or even asteroidal belts like ours for that matter. From here we can ask whether these solar wannabes also have 11-year sunspot cycles and all the other detailed features of our Sun. The important point is that there are a *lot* of Sun-like stars that Earth would be equally proud to orbit!

111   How will the Sun change in the next 10 billion years?

In a research paper published in the November 20, 1993, *Astrophysical Journal,* astronomers I.-Juliana Sackmann, Arnold Boothroyd, and Kathleen Kraemer, from Caltech, University of Toronto, and Boston University, used the best stellar evolution models now available to answer this exact question. They "watched" as the Sun started out 4.6 billion years (byrs) ago with a luminosity of 0.70 of today's value and brightened to 2.2 times its current value by 6.5 byrs hence. A luminosity of 1.1 times the present value was reached in 1.1 byrs, and 1.4 times its present power output some 3.5 byrs from now. This means that in 1.1 byrs, Earth will experience a permanent "moist greenhouse effect" as the oceans begin to evaporate into the atmosphere, and by 3.5 byrs, a "runaway greenhouse effect" similar to what Venus is enjoying will be in place. As the Sun evolves into a red giant, it grows to a luminosity of 2,300 times its present value and a size of 150 times its current extent, shedding about 27 percent of its mass and engulfing/incinerating the planet Mercury. It continues to evolve up the so-called asymptotic giant branch and experiences at least four "thermal pulses." After the first one, the Sun has swollen to 213 times its present size, but at a mass of only 0.6 of its current mass, the orbits of the planets have crept outward from their present distances. Venus by that time would be located about 1.22 times farther from the Sun than where Earth is now. Earth will have moved to where Mars is today, so neither of these planets would be engulfed by the Sun. The Sun will reach a peak luminosity of 5,300 times its current rate after the fourth thermal pulse.

112   How large is the Sun's core?

Theoretical models and actual measurements by the SOHO MIDI helioseismology experiment suggest that the nuclear furnace that powers the Sun occupies a volume about 10 percent of the diameter of the Sun. This works out to ten times the diameter of Earth. It is defined as the region where temperatures are high enough to allow thermonuclear fusion to occur at temperatures over one million degrees K. When the Sun becomes a red giant, this core will be compacted to about the size of Earth (4,100 miles in radius) and will be a white dwarf.

113   How long does it take light to get out from the inside of the Sun?

Some textbooks refer to "hundreds of thousands of years" or even "several million years," depending on what physical circumstances are assumed. Also, the interior of the Sun is not at constant density so that the steps taken by light rays in the outer half of the Sun are much larger than in the deep interior where the densities are highest. Most astronomers are not too interested in this number, and forgo trying to pin it down exactly because it does not impact any phenomena we measure. These estimates show that the emission of light at the surface can lag behind the production of light at the core by something like 100,000 years or more. The point of all this is that it takes a *long* time for light to leave the Sun's interior. A more precise answer may not be of much scientific interest right now.

CHAPTER 4

The Solar System—Living in the 'Hood

It was, once upon a time, all that existed in the universe. Even the now-distant stars crowded the hearth-fires of Sol from a distance not much beyond Saturn. As we gained greater telescopic sight, our little family of seven celestial bodies were joined by three more worlds traveling the distant tracts of space. In the span of only a few generations, this cozy little family has swelled to over 100,000 motes of rock and ice—some no larger than a house. A few of these have become so well known to us that pictures of them captured by passing satellites raise no more astonishment than a tourist's photo of the Grand Canyon or Machu Picchu. A murky photo of Mars taken in 1965 is replaced by thousands from a planetary orbiting platform as it methodically captures every boulder and hillock larger than a few dozen yards across. We have leaped beyond the confines of our planet, and now see our neighbors as less of a mystery and more of a familiar spot we have visited before. In the centuries to follow, we will swarm its deepest recesses with technologies and intentions we can scarcely imagine today. Humans will reach Mars, walk its sandy ochre deserts and permafrost, and perhaps fashion a few outposts beyond this world in the farther reaches of our solar system. But the binding forces of our humanity, and our needs for Mother Earth, will pull us back home. The unimaginably lonely voids that stretch beyond Pluto will never be a welcoming vista for us.

◑ ● ◐

114 Does Mercury rotate?

It is one of the hardest planets to spot in the sky because it never strays more than a few dozen degrees from the Sun. It often gets lost in the clutter of the eastern or western horizon. In 1889, astronomer Giovanni Schiaparelli's

(1835–1910) telescopic observations of dark features on the planet's surface seemed to prove that its day was 88 Earth-days long, and that is exactly what textbooks continued to say until around 1960. By then, radar pulses reflected from the surface were successfully detected, which led Cornell University astronomers Gordon Pettengill and Rolf Dyce to conclude in 1965 that Mercury rotated once every 59.3 days, give or take two days. This has been steadily improved to the current value of 58 days and 15.6 hours, give or take 15 minutes. A little-known, but fascinating, phenomenon on the Mercurian surface is that during certain times of its year, the Sun rises twice near the poles. This happens when Mercury is closest to the Sun at perihelion and the angular speed of the Sun across the sky is about the same as the daily sky rotation speed. From polar latitudes, the Sun can execute a retrograde motion that doubles back on itself, causing a double sunrise, or even a double sunset. The next opportunity for this to occur will be in December 2003.

115 Are there ice caps on Mercury?

Although the equatorial temperatures are near 800 F, because the rotation axis of Mercury is nearly perpendicular to its orbit around the Sun, there are certain craters near Mercury's north pole where the Sun never shines, and where temperatures as low as −235 F can occur. In 1992, astronomers Martin Slade, Bryan Butler, and Duane Muhlman at the Jet Propulsion Laboratory and Caltech reported the first full-disk radio image of Mercury using the Goldstone 70-meter radio telescope. As Figure 9 shows, they found that some of the craters had radar reflections as bright as those seen on Mars, and similar to the ices on the satellites of the outer planets. The ice must be mixed with the "topsoil" like permafrost. The best explanation for their existence involves water-rich comets that crash into Mercury. Some of the water molecules eventually find their way to the cold traps in the shadowed craters. Why is this important? Because it tells us that one of the most important ingredients for life is very hardy on planetary surfaces. Also, if for some unimaginable reason you would want to land humans on Mercury, this is the best possible place to put them to save on bringing water with you. Mercury's ice could be purified into drinkable water at very low cost using a solar-powered still. In the future we will do more surveys of this potential water deposit, and perhaps have a robot lander visit this area and do some digging! Can you imagine what a glass of imported Mercury water might cost?

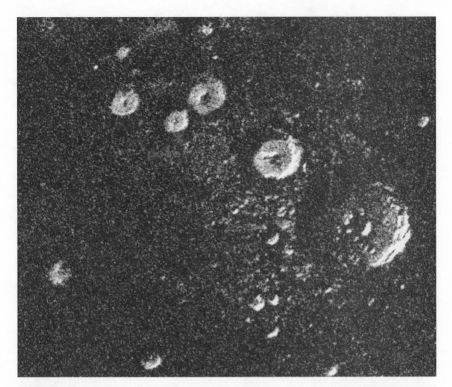

FIGURE 9 Mercury ice caps? Radar pulses reflected from the polar regions of Mercury and detected by the Arecibo Telescope reveal bright regions in craters that may be surface or sub-surface traces of water ice bound up in the soil. Located in the perpetually shadowed regions of these craters, comet ice may have accumulated there for billions of years. (CREDIT: JOHN HARMON, ARECIBO OBSERVATORY)

116 Who discovered Venus?

No one knows. Venus as a wandering star was known to the ancient Egyptians and Babylonians based on their writings, but we don't know the name of the individual who discovered it. The earliest cuneiform inscriptions about the Star of Ishtar come from the Venus Tablets of Ammizaduga, which were produced by the Babylonians around 1650 B.C. Ancient Egyptians also knew about Venus, which they called Hathor's Star, and often associated it with Osiris. But their mentions of Venus seem to appear much later in their history, during the Greek Ptolemaic era. We have the same problem with the Sun and Moon. We don't know who "discovered" them either.

117   Why don't Venus and Mercury transit the Sun every year?

The orbits of these two planets are inclined to Earth's orbit, and their sizes are so small that the timing has to be exactly right for them to pass through the ecliptic plane exactly at the time when the ascending/descending node is between the Sun and Earth. For Mercury, this happens every five to ten years. For Venus this is a lot less common. The May 7, 2003, transit of the Sun by Mercury was observed by the NASA TRACE satellite, which returned some excellent images and movies of the event. The upcoming Venus Transit on June 8, 2004, will be much watched because this will be the first time we will see it since its last appearance on December 6, 1882. In Figure 10 you can see

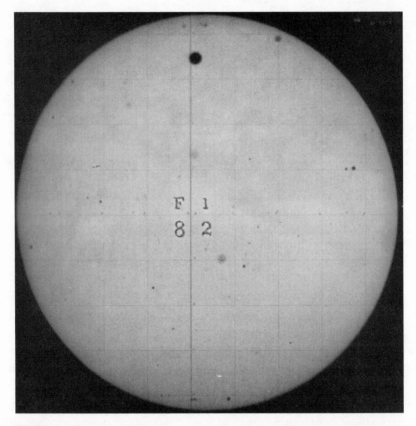

FIGURE 10 The Transit of Venus ca. 1882. This U.S. Naval Observatory photograph captured Venus as it passed across the face of the Sun. It was an international event, and thousands of people lined the streets of New York City to peer through street-corner amateur telescopes, or smoked glass, to participate in this rare event. (CREDIT: U.S. NAVAL OB-SERVATORY LIBRARY)

a U.S. Naval Observatory photograph of this historic event. If you want to participate in a NASA-led, North American observing project with students and amateur astronomers, visit the NASA, Sun-Earth Connection Education Forum website (http://sunearth.gsfc.nasa.gov) and click on the Big Events button for 2004. It will be only the fifth such transit ever witnessed by humans since it was first anticipated and observed by Jeremiah Horrocks (1618–1641) on December 6, 1631.

118   Do the phases of Venus have to do with a shadow cast by Mercury?

Just as the Moon goes through its monthly changes from new moon to full moon, the planets Venus and Mercury also go through similar illumination changes as viewed from Earth. These changes have nothing to do with the shadow of one planet moving across the other. The darkest part of Mercury's shadow cone has its vertex about 114,000 miles from Mercury and doesn't reach Venus at all. The phases are caused by the fact that we see the night side of Venus itself as it is turned toward Earth in varying degrees along its orbit. We see an almost fully illuminated Venus when it is nearest the Sun as seen from our sky. When Venus is nearest Earth in its orbit, we see Venus as a thin crescent.

119   Why did Mars run dry?

We think that the destiny of Mars was sealed by something as simple as its smaller mass, which forced a rapid evolution of its atmosphere and interior. It may have been an entirely Earth-like world three billion years ago. Eons later, it has become a cold planet that has largely lost its atmosphere. Yet in its exposed, ancient crust we see written a fascinating story of rivers that once flowed, and vast, but shallow, oceans that once graced its rusty-red lowlands. In Figure 11, taken by the NASA Mars Global Surveyor satellite, we can see these gullies carved into valley walls and running down-slope. Since the earliest telescopic viewing of Mars, its north polar cap has given us tantalizing hints that water-ice (not just carbon dioxide–ice) may be present at the surface in vast, exposed deposits. Intense solar ultraviolet radiation breaks water molecules into free hydrogen and oxygen. Released from its molecular prison, oxygen flows into the atmosphere and is quickly lost by incessant collisions with solar wind particles at a rate of something like 30,000 tons per year. On Earth, this loss is slowed to only a few hundred tons per year. Part of the vast difference is in the masses of the planets, but it is also a matter of a much less obvious ingredient: their magnetism. Earth's magnetic field

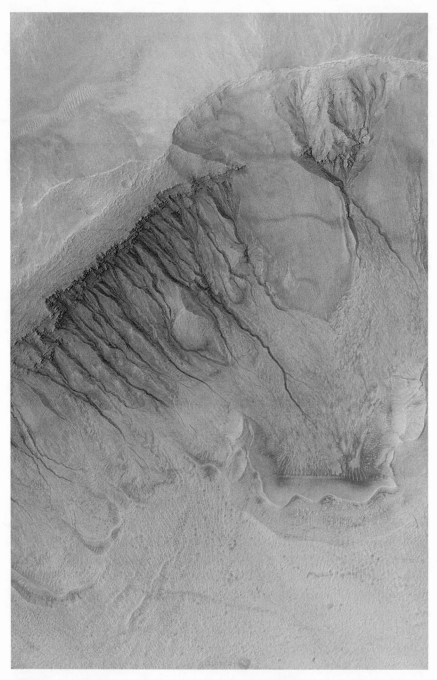

FIGURE 11 A watering hole for Martians? Gullies seen on the canyon walls of Nirgal Valis may be subsurface water draining down-slope, causing mile-long channels. (CREDIT: NASA/JPL/MALIN SPACE SCIENCE SYSTEMS)

deflects much of the solar wind away from Earth's atmosphere like an invisible umbrella that prevents the wholesale erosion of our atmosphere. Even the feeble crustal magnetism of Mars seems just enough to slow down the loss of its atmosphere through solar wind erosion.

120 What do you think about the Face on Mars?

Not much. Wind-blown sand can produce lots of amazing shapes over time and we know that Mars has plenty of strong winds and exposed rocks. I am actually amazed that geologists looking at Mars haven't seen even stranger shapes. Of course there were the famous Martian canals in the 20th century, but these vanished once we got to Mars and took pictures firsthand using unmanned spacecraft. As for the Martian Face, well, the Mars Global Surveyor re-imaged that region on April 5, 1998, and on April 8, 2001, with two-yard resolution, only to see it as a mesa that had been shaped, as expected, by sandstorms. In Figure 12, each feature in the Face can be traced back to a geologic feature in the mesa, when allowance is made for the different Sun angles in the satellite photographs. Many people made tens of thousands of dollars in book royalties, lectures, and radio interviews hyping NASA's Viking orbiter "discovery" of a mysterious face on Mars or a shape resembling one. This picture became a mainstay of many fringe discussions of space aliens and the like. It was even the central story in the otherwise entertaining year-2000

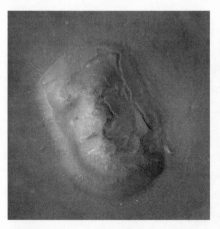

FIGURE 12 Face on Mars. A recent close-up image (a) by the Mars Orbiter Surveyor in 2001 (resolution 7 feet per pixel) looks a lot different than (b) the suggestive image captured by the Viking Mission in 1976. It sure looks like a mesa to me. (Credit: NASA/JPL/Malin Space Science Systems)

movie *Mission to Mars*. Not a single respectable astronomer made so much as a dime from this hoopla, except perhaps by being invited by various groups to give public talks about why the whole idea was silly in the first place. Astronomers are often asked to be the "clean up squad" after some silly idea makes a big media splash.

121 What do we know about the famous Mars Meteorite and life on Mars?

Of the 22,000 meteorites that have been recovered on Earth, only 28 have been confirmed through isotopic studies to be probable Mars meteorites. The most famous of these is ALH84001, which made front-page news on August 7, 1996, as a source of fossil Martian microorganisms. Since then, many scientists have studied this five-pound stone with the most sophisticated chemical and microscopic techniques you can imagine. Also, NASA has created an entire new division of Astrobiology to aggressively study extraterrestrial and extreme-terrestrial life. A major expansion of the Arctic meteorite recovery program was also supported by NASA with a $1.6 million grant to scientists at Case Western Reserve University. Some 11,500 meteorites have already been recovered from Antarctic ice fields by these science teams. On August 2, 2002, NASA announced that further evidence had been found that the meteorite contained fossils of Martian organisms. Kathie Thomas-Keprta, an astrobiologist at NASA's Johnson Space Center (JSC) in Houston and the lead researcher on the study, and her colleagues had been studying the nanometer-sized magnetic bodies found in the meteorite and concluded that a quarter of them could only have been produced by a process similar to terrestrial magnetotactic bacteria. Rather than a random collection of inorganically produced magnetite, these were very pure crystals of the kind that Earth bacteria create in order to use Earth's magnetic field to navigate. The fact that NASA satellites have also confirmed that Mars had a strong magnetic field in the past seems to go along with the idea that Martian bacteria evolved to use it as a navigation tool. But is this proof of life on Mars? Perhaps not. Researchers David Barber of the University of Greenwich and Ed Scott of the University of Hawaii don't think those tiny crystals of iron oxide in ALH84001 were formed by organisms at all. The planes of oxygen atoms in the magnetite crystals are aligned with those in the surrounding carbonate crystal. This proves, they contend, that the magnetite crystals could not have formed elsewhere and then been deposited in the carbonate. They must have formed exactly where they are observed today in the carbonate. This observation is counter to what the NASA-led team has argued. So, even now there still isn't a clear consensus, and that is exactly what we need to prove the case for life on Mars. We may well have to actually go to Mars to find out, but then, that was the plan all along!

122   What is the best possible place to search for fossils on Mars?

In an exciting press conference in June 2000, astronomer Michael Malin of Malin Space Science Systems in San Diego announced that NASA's Mars Global Surveyor spacecraft had found geographic features in many locations on Mars created by flowing water. Such characteristic landforms were found at approximately 150 locations, including the inside of craters in about 25,000 Surveyor images of Mars. They were supposedly created by groundwater as it penetrated to the surface and flowed over it. Although the images didn't show actual water, the discovered V-shaped, eroded channels as well as their branching patterns closely resemble rivers that run through mountains on Earth. The water supply is believed to be about 300 to 1,300 feet below the surface, and limited to specific regions of Mars. Most of the wet spots were found on the walls or central peaks of impact craters. Others turned up on the walls of distinctive pits in the south Polar Regions. Still more were discovered on the walls of two major Martian channel systems, Nirgal Vallis and Dao Vallis. The flows that came down each gully may have had a volume of water enough to fill seven community-sized swimming pools, or sustain 100 households for a month.

123   At what distance would Mars be eclipsed by Earth's shadow?

This would never happen. Earth subtends the same angular size as the Sun about 930,000 miles from Earth. At the distance of Mars, Earth's greatest possible angular diameter is about 48 arc seconds, while the Sun subtends 1400 arc seconds. Earth would not produce a full eclipse of the Sun but would be seen as a small black spot transiting the bright disk of the Sun. In fact, as seen from Mars, the transit of the Sun by Earth would look very much like the upcoming transit of Venus on June 8, 2004. On November 10, 2084, Earth will pass across the disk of the Sun as viewed from Mars. We can only hope that there will be humans there to see it.

124   Where can I get information about terraforming Mars?

How would you like to alter another planet to mimic Earth's shirt-sleeve environment? Since the 1970's, some scientists and engineers have imagined how such "terraforming" projects could be undertaken to eventually make

Mars habitable. The issue of terraforming has very potent moral, political, and scientific aspects to it. Many people have spent serious time thinking about the ramifications of taking over another planet and altering it to suit humans. Personally, I think it is virtually impossible to alter another planet in this way. Mars is a leaky bag so far as creating a new atmosphere is concerned. It loses its atmosphere at a rate of about 30,000 tons per year. We would have to not only exceed this rate but also greatly surpass it to have a chance for a dense-enough atmosphere to form in a century or so. Any longer time, and this project would have a payoff too far in the future to be of interest to us.

125   Are the Viking landers still working?

After returning thousands of pictures to Earth since 1976 like the one in Plate 6, they both shut down many years ago when their batteries failed. NASA had planned that the Viking mission would continue for 90 days after landing. Viking Orbiter 1 functioned until July 25, 1978, while Viking Orbiter 2 continued for four years and 1,489 orbits of Mars, concluding its mission August 7, 1980. Because of the variations in available sunlight, both landers were powered by radioisotope thermoelectric generators (RTGs)—devices that create electricity from heat given off by the natural decay of a few grams of plutonium. That power source allowed long-term science investigations that otherwise would not have been possible. The last data from Viking Lander 2, located in Utopia Planitia, arrived at Earth on April 11, 1980. Viking Lander 1 made its final transmission to Earth November 11, 1982, from its picturesque location in Chryse Planitia.

126   Can Mars hold an atmosphere as thick as Earth's?

There are a number of aspects to Mars that would make holding onto an atmosphere very difficult over the long term. One of the major difficulties is that its mass is so low it can't gravitationally hold onto an atmosphere that would be heated by the Sun at this distance. Like a slowly boiling pot, the atmosphere eventually leaks away. Also, without a magnetic field, the solar wind interacts directly with the atmosphere and erodes about 30,000 tons of atmosphere yearly into space. This also happens on Earth, but its higher mass and stronger magnetic field make the process very slow. We also have volcanic activity and a biosphere, which help to generate atmospheric constituents. On Mars, there has not been volcanic activity for billions of years, so the loss of its perhaps once Earth-like atmosphere rich in carbon dioxide was never replenished volcanically, or biologically. The atmosphere may also have simply been chemically absorbed by the planet into its surface rocks. Another possibility is that Mars is so close to the asteroid belt that major impacts by large bodies have explosively ejected large quantities of the Martian atmosphere into space.

127 When will we return to Jupiter to study its moons up close?

The last time we were able to study this planet up close was when the NASA Galileo spacecraft went into orbit in December 1995. For the next eight years, it followed a complex series of orbits that brought it near its four largest Galilean satellites, Io, Europa, Ganymede, and Callisto, for spectacular close-ups. NASA has been given the green light to develop, within the next ten years, a nuclear propulsion system called Prometheus that will allow a spacecraft called the Jupiter Icy Moons Mission to travel to Jupiter and thoroughly explore its large Galilean moons, Europa, Ganymede, and Callisto. This support came rather suddenly when President George W. Bush signed the $3 billion Nuclear Propulsion Initiative for the 2004 U.S. budget. This will allocate money over five years to design and build a lightweight 250-kilowatt nuclear reactor that will be used to power the satellite. Currently, outer solar system spacecraft rely on radioisotope thermoelectric generators (RTGs) that develop electricity by using the heat from radioactive decays of low-grade plutonium to produce a few hundred watts of electricity. With 250 kilowatts of "stadium lighting," many complex power-hungry experiments can be operated, along with an ion drive with much higher thrust than the Deep Space 1 mission.

128 Jupiter produces its own energy, so why can't we think of it as a star?

The way that astronomers define "star" has to do with whether an object is fusing one element into another to produce light. This requires central temperatures in excess of several million degrees to begin the ignition of hydrogen fusion, which defines the main energy source for Main Sequence stars. The brown dwarf called Gliese 229B (GL229B) is a small companion to the cool red star Gliese 229, located 19 light-years from Earth in the constellation Lepus. Estimated to be 20 to 50 times the mass of Jupiter, GL229B is too massive and hot to be classified as a planet as we know it, but too small and cool to shine like a star. This object was discovered in 1994 with the 200-inch Hale Telescope at Palomar Mountain by a team of Johns Hopkins and Caltech astronomers led by Shrinivas Kulkarni. The dwarf has the spectral fingerprint of the planet Jupiter—an abundance of methane. Methane is not seen in ordinary stars, but it is present in Jupiter and other giant gaseous planets in our solar system. Hubble data obtained in 1995 show that the object is far dimmer, cooler (no more than 1,300 degrees Fahrenheit), and less massive than previously reported brown dwarf candidates, which are all near the theoretical limit (8 percent the mass of our Sun) where a star has enough mass to sustain nuclear fusion.

129 What is the surface of Titan like?

Its density is about one-third that of water (1.9 gm/cc) so this suggests the moon is rich in rocks that contain lots of hydrogen. The material is probably similar to what astronomers have studied in carbon-rich, chondritic meteors. Titan has an atmosphere whose pressure at the surface is an incredible 1.5 times that of Earth's atmosphere at sea level. The temperature at the surface is estimated to be −290 F. The atmosphere has been spectroscopically studied, and it seems to be a witches' brew of molecules you might expect to find at your local gasoline pump. Besides nitrogen, methane, and hydrogen, you can find propane, acetylene, ethylene, hydrogen cyanide, and cyanoacetylene. Its surface receives about one percent the sunlight that Earth does, and it is expected from chemical models that ethane probably condenses out of the atmosphere like rain, and has formed an ocean perhaps one mile deep with a sediment of solid acetylene up to 600 feet thick at its bottom. In reality, given the overabundance of theoretical models, the oceans are probably little more than large lakes or puddles spread across the surface. Titan is not a habitable place even with its solid surface and thick atmosphere, but it shows that Mother Nature likes to build some very large factories for cooking organic molecules under almost impossible conditions. Just think what she cooked up on Earth under warmer conditions about 4.5 billion years ago. The Hubble Space Telescope has taken intriguing images of the surface of Titan that show hot and cold regions that are perhaps continents surrounded by oceans of some kind (see Plate 7). On Christmas Day 2004, the Cassini spacecraft will have deposited the Huygens probe on the surface and we will have an even better idea of what this fascinating world is really like.

130 How thick are Saturn's rings?

Before the Voyager 2 spacecraft flew by Saturn on August 26, 1981, astronomers deduced that, because Saturn's rings disappeared when seen edge on, the ring system must be less than 25 miles thick. In 1973, radar pulses were bounced off of the rings, and showed that the ring particles were not much more than a few yards across. Studies of the absorption of light from distant stars eclipsed by the rings were eventually able to detect starlight absorption by ice, so from all of this we deduced that the rings were ice-coated, very reflective small bodies that orbit Saturn by the hundreds of billions. Voyager 2 flew by the rings in the 1980's and showed them to be not eight to ten entities, but thousands of individual ringlets. Some rings were only a few miles wide. The thickness of the ring plane is now thought to be less than 300 feet! If you collapsed all of the vertical spaces between the ring particles, they would form a disk only six feet thick, but extending over 180,000 miles from Saturn, making these rings the thinnest objects in the solar system. We are lucky to be here to see them today, because computer calculations predict that they will probably disappear in another few hundred million years or so.

131   What makes Uranus and Neptune different colors?

They are essentially identical planets with nearly the same radii (15,500 miles), mass (16 times Earth), density (1.5 grams/cc), and rotation (16 hours), so how is it that Uranus has a distinctly blue-green hue while Neptune is decidedly blue? Both planets are made of hydrogen and helium (nearly 100 percent) with traces of methane gas. For Uranus, this gas absorbs the orange and red sunlight components and scatters the blue-green component back to us to give it its color. Being closer to the Sun, the atmosphere has a higher haze layer, rich in hydrocarbon "smog." This renders Uranus almost featureless. Neptune also has a haze layer, but it is located deeper in the atmosphere. This layer absorbs even more of the sunlight, leaving a bluish color to it. Because the haze layer is deeper, we see more clouds on Neptune than on Uranus.

132   What is Pluto like?

Our understanding of this distant world has improved considerably since the 1960's and 70's when all that textbooks could say was that it was a cold planet, with a diameter between 4,700 and 3,600 miles, a distance of 39.5 AU, and a rotation period of 6.4 days. Telescopic images showed merely a faint star shifting position in an otherwise unremarkable star field. In 1978, a satellite to Pluto was detected by astronomer James Christy at the U.S. Naval Observatory, and quickly named Charon. This satellite is 750 miles in diameter and has a density of about one-quarter that of water. It orbits Pluto in a synchronous manner just as our Moon does in its journey around Earth, but Charon's distance is only an astonishing 10,000 miles. The satellite's orbit and period allowed astronomers to precisely determine a mass and size for Pluto for the first time. It has a diameter of 1,413 miles and a mass of 1/450 Earth, for a density of only half that of water! As seen from the surface of Pluto, Charon would appear eighteen times bigger than our Moon, and it would travel across the sky at a speed of two degrees per hour, some four times faster than our moon. It would be an impressive sight, to say the least.

Pluto is a planet barely as large as some of the satellites of Jupiter and Saturn and has a composition that must be very rich in water and water ices. Its density is completely consistent with its having been formed in the outer reaches of the solar nebula, according to the best chemical models that astronomers have of this ancient time. At an expected surface temperature of −382 F, water is certainly in ice form. This is cold enough that methane gas can condense on the surface as frost. Alan Stern at the Southwest Research Institute in Boulder, Colorado, and his colleagues used the Hubble Space Telescope and presented some tantalizing images of Pluto in March 1996, such as the one in Figure 13. The images revealed a surface mottled by dark and bright "continents" perhaps as regions where reflective ices covered a darker, organic-rich crust. By 1992, astronomers were able to detect frozen ices on its surface, and we now know its composition as 97 percent nitrogen,

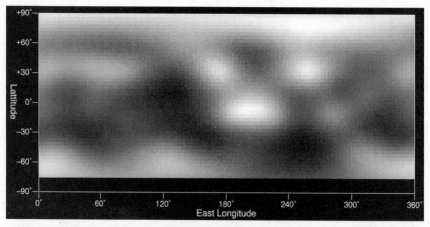

FIGURE 13 Pluto surface assembled from four separate images of Pluto's disk taken with the European Space Agency's (ESA) Faint Object Camera (FOC) aboard NASA's Hubble Space Telescope. (CREDIT: ALAN STERN [SOUTHWEST RESEARCH INSTITUTE], MARC BUIE [LOWELL OBSERVATORY], NASA, AND ESA)

1 to 2 percent carbon monoxide, and 1 to 2 percent methane. Its atmosphere can be detected when the planet passes in front of a star, but it is only present when Pluto is close enough to the Sun to heat its surface ices to vapor. This has happened since June 1988 and was observed by a team of astronomers who were able to detect a very dilute atmosphere of methane and nitrogen. But this time span of a Plutonic atmosphere is rapidly drawing to a close as Pluto moves along its orbit farther from the Sun. By 2015 it will be cold enough that its atmosphere will once again freeze onto its surface. The proposed NASA Pluto Express mission may well not have an atmosphere to study if its launch is delayed beyond 2005, which seems likely.

133 What is the difference between asteroids, comets, and meteorites?

Asteroids are large, rocky bodies from yards to miles in size often either stony or iron-nickel in composition, with some rich in carbon compounds and classed as carbonaceous chondrites. Most that strike Earth seem to be from the asteroid belt, with orbits perturbed by collisions among themselves, or distant orbit-changing encounters with Jupiter or Mars. Comets are large mile-sized bodies rich in ices along with dust grains, gravel, and small rocky bodies embedded in the ice. When the ices evaporate, they grow tails and we see them as comets. They leave behind in their orbits gravel and rocky debris that Earth may encounter from time to time as orbits cross. We see these encounters as meteor showers. Meteorites are small rocky bodies, from micron-sized up to perhaps a meter across, that impact Earth after a brilliant atmospheric display. They can be fragments of material chipped off of asteroids that collided. They can be rocky cometary debris, or they can be material in interplanetary space that over 4.5 billion years just never got incorporated into cometary bodies, planets, or asteroids. A meteor is a small body, usually the size of a pebble or sand grain, that enters the atmosphere and burns up, but some of the larger ones a foot or more across can reach the ground and are called meteorites when their fragments are recovered.

134 What are most asteroids made from?

Asteroids are made from ices and rocks, or probably combinations of these ingredients. The rocks can be iron/nickel, carbon compound–rich, or stony materials similar to terrestrial crustal rocks. It is very important to understand what asteroids are made from, because one day we may be asked to take action against one of them to avoid being hit. The current investigations seem to suggest there are some looming problems ahead. The biggest problem is that a surprisingly high percentage of asteroids have such low densities that they should not be thought of as solid objects at all, even though they may have craters on their surfaces. According to Dan Britt (University of Tennessee) and Guy Consolmagno (Vatican Observatory), the densities of many (but not all) asteroids are substantially below their most likely meteorite analogues. They are probably gravitationally bound rubble piles, some of which have more empty space in them than solid material. Asteroids such as Deimos, 253 Mathilde, 45 Eugenia, and 16 Psyche have average densities substantially below the least-dense meteorites ever found. A rocket striking one of these asteroids would be like a hammer hitting a pillow.

135   Is there an asteroid that will hit Earth in the next few years?

There are no known bodies on a direct collision course. As of June 17, 2003, the International Astronomical Union's Minor Planets Center has a list of 512 objects with closest Earth approaches less than five million miles. One of these, called 1950 DA, seems to be headed for a rendezvous with Earth on March 16, 2880. The best orbit models from the data we have suggest that it could come within 30,000 miles of Earth. No plans are being made to intercept it, because with our present technology we might make the situation worse. Most of these Near-Earth Objects (NEOs) have been discovered since 1998, when careful surveys for them were started by NASA with the help of the U.S. Space Command. These bodies are several miles across and would be devastating if they hit Earth. What is a bit more nerve-wracking is that we cannot easily see the smaller chunks of rock less than a half mile across that are far more numerous. On August 10, 1972, a famous daytime "fireball" skipped across the Grand Teton Mountains in Wyoming before returning harmlessly back to space over Canada. This object was estimated to be 200 yards across. If it had landed on one of thousands of towns or cities, it would have left a crater a half mile across. Needless to say, the inhabitants would have had absolutely no useful warning, and nowhere to escape. On June 17, 2002, a large 40-foot body called 2002MN was detected as part of the Spaceguard Project. They calculated its orbit and discovered that it had made its closest approach to Earth of 75,000 miles about three days earlier. If it had collided with Earth, it would probably have detonated in the atmosphere with a yield of 10 megatons. It is estimated that tens of thousands of objects down to a size of a few meters remain to be found, and any one of them could pay us an unexpected and potentially devastating visit. But objects don't have to actually impact the ground to be a problem.

Thanks to the global infrasound network set up by the military to monitor nuclear explosions, about one atmospheric detonation is detected every month with the explosive yield similar to the Hiroshima bomb (15,000 tons of TNT). On April 23, 2001, researchers at Los Alamos National Laboratory heard a meteor explosion over the Pacific Ocean with a yield of a whopping 6,000 tons of TNT. The object was estimated to be about 12 feet in diameter. One of these Hiroshima-level airbursts occurred on June 6, 2002, over the Mediterranean. Experts believe that if it had come down two hours earlier over Kashmir, it might have triggered a panic leading to a nuclear exchange between India and Pakistan. Table 1 is a list of the most famous detonations recorded so far. It is not a complete list, because it doesn't include the large number of smaller detonations heard each month. Until we can mount a very aggressive and systematic search for these insidious natural hazards, we will remain very much at their complete mercy.

136   How did asteroid Ida end up with a companion moonlet?

The companion asteroid called Dactyl is a small scrap of rock just under a mile across discovered in 1993 during a fly-by of the Galileo spacecraft as it was headed toward Jupiter. We now know of six "binary asteroids" in the solar system after only a few years of active searching, so whatever the mechanism that causes them, it isn't a rare event. In 2001, computer simulations by Patrick Michel (Observatoire de la Côte d'Azur) and his colleagues have shown that, when asteroids collide, some do so at slow enough speeds that the asteroids are shattered but not completely dissipated. Instead they reform as new asteroids, and in many instances, unincorporated fragments become satellites. Miranda, a satellite of Neptune, is thought to be a previously shattered and re-formed body.

137   Do asteroids have dust on them?

Between October 25 and 26, 2000, NASA spacecraft operators brought the NEAR spacecraft to an orbit within three miles of the surface of the asteroid Eros. The data revealed a large number of small boulders, lots of very large craters, and a lack of small craters. Thousands of small boulders 10 to 12 feet across were found all across the surface of Eros, but very few small features on the surface, suggesting that the surface isn't solid. According to Dr. Joseph

Veverka of Cornell University, who headed the imaging team, Eros has a regolith about as deep as a ten-story building. It is probably made of a combination of dust and fine particles produced by billions of years of thermal
expansion and contraction, and micrometeorite bombardment. Also, there
are very smooth and flat "ponds" scattered across the asteroid that look as
though the regolith has been sorted into regions with very fine particles. This
may also have been produced by the seismic shaking of the regolith. On February 12, 2001, NEAR actually landed on the surface of Eros and recorded
surface details as small as four inches, as shown in Figure 14. During its de-
scent, it studied flat-bottomed craters probably filled with electrostatically
charged dust. It also examined boulders that were eroding and shedding material into aprons, but with no sign of a plausible mechanism around to do
this other than what has become the standard catch-all explanation: micro-
meteoroid impacts.

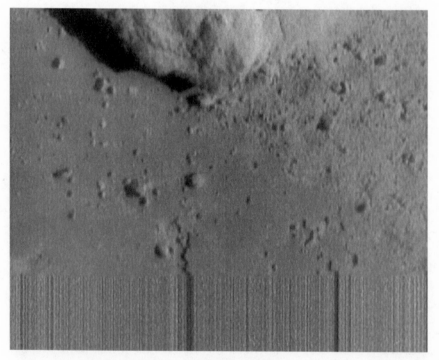

FIGURE 14 Asteroid Eros close-up. This is the last image returned by the NEAR
satellite before it landed on 433 Eros. Taken from an altitude of 420 feet,
this image shows a large rock 12 feet across at the top. (CREDIT:
NASA/NEAR/NSSDC)

138 Why are asteroids made from rocks instead of fluffy things since they were never inside anything to remelt them?

Most of the primordial material of the infant solar system has been processed in large bodies that apparently broke up from collisions during the planet-building era. Interstellar dust grains are chemically sticky, snowflake-like objects. But once you get them to grow into inch-sized clumps, new grains impact the older structures and compress them somewhat, like a baseball compressing the fabric in a catcher's mitt. Further collisions cause these small, sticky rocks to grow into large asteroids, but they are not very solid objects at all. Asteroids like Mathilda have the density of loose gravel and are not solid like a surface rock on Earth. Once asteroids get to be larger than about 200 miles in diameter, they sag under their own weight to form round bodies, with central densities higher than surface densities. At a size of about 1,000 to 2,000 miles across, trapped heat from the decay of radioactive elements or heat of formation can partially liquefy the interior, and we get the kind of mass segregation that leads to stony and iron-nickel meteorites. Astronomers know of at least five distinct families of asteroids, thought to be the remains of individual planetessimals somewhat larger than Ceres (280 miles in diameter) that apparently collided in the asteroid belt to create the fragments we see. At least one of these bodies was big enough to produce the highly differentiated meteorites we recover on Earth. That's why comet studies are so important, because they represent the earliest collections of matter prior to the differentiation stage. Carbonaceous chrondrites are also a very early, lightly processed form of primordial "rock," containing volatile, trapped gases and carbon compounds . . . many that are very heat sensitive.

139 How are comets formed?

We don't really know. There is no good way to "date" comet material, especially since we don't have any good samples of their material from the scraps of comet dust that we can collect in the stratosphere. They may predate the formation of our solar system, but we don't know whether they were formed inside ancient interstellar clouds, or in the disk of gas and dust that once orbited the infant Sun. They do represent very primitive matter with chemical compositions similar to gases found in many interstellar clouds today. The NASA Stardust mission will arrive at comet Wild 2 on January 2, 2004, and capture some of its tail material, then alter course and return to Earth where the sample of comet material will be recovered. Soon after the sample arrives back on Earth on January 15, 2006, we will know the age and chemistry of at least one comet.

140   Why don't the ices in comets just sublimate away?

Sublimation only happens if the ice is near its freezing point so that there is still considerable internal energy at the ice surface to cause molecules of the ice to "boil off" from its surface. For comets and ices on the lunar surface in shadow, the temperature is well below the freezing point of the ice, so there is little internal heat available to boil off the molecules. The result is that ice molecules remain where they are, and in the case of the Moon, any new ice molecules in the atmosphere get stuck to this core of ice, which then acts like a "cold trap" pumping ice molecules out of the atmosphere. This same principle is used in vacuum pumps in Earth labs. As they get heated by the Sun, the surface atoms get warm enough that they can start to evaporate.

141   Do all comets come from outside our solar system?

There are actually three types of comets astronomers can distinguish once they have made enough measurements to compute their orbits. Short-period comets have elliptical orbits that bring them into the inner solar system (inside the orbit of Mars) every 200 years or less. Some of the famous short-period comets include Halley's (76 years) and Encke (3.3 years). These comets seem to have come from inside the orbit of Neptune, and there are enough of them that there must be quite a few of these comets-in-waiting ready to get bumped into an orbit that takes them deeper into the solar system. The next class is the long-period comet. These have very steep elliptical orbits that take them far beyond the orbit of Pluto, and it is believed that this is where the reservoir for these comets occurs—possibly the Kuiper Belt. Orbit periods can be thousands of years long and are at the limits of our ability to detect because they are rare and because their orbits are so steep that it takes many separate measurements to really tie them down to such extremely elliptical paths. The last category could be informally called the interstellars, which would have speeds so fast they must be on orbits that begin and end in the distant Oort Cloud, or perhaps even interstellar space. Any object seen with a speed greater than 90,000 miles per hour cannot be gravitationally "bound" by the Sun. Many of these are believed to have been long-period comets that were sling-slotted by a close encounter with Jupiter into the orbits we now find them. So far, astronomers have not detected any of these visitors from interstellar space.

142   How fast do asteroids and comets rotate?

Observations of Halley's Comet by the Vega and Ghioto spacecraft, as well as ground-based studies in the 1980's, showed two nuclear rotation periods of this potato-shaped comet nucleus: a 2.4-day period, and a 7.4-day period. Studies of asteroids such as Eros and Icarus along with 124 others show that they spin at many different rates from 226 hours (minor planet 1982FT) to as little as six minutes (minor planet 1999 TY2).

143 How many moons do each of the planets in our solar system have?

Since 1998, there has been a dramatic increase in the number of moons detected around distant planets. For the inner planets, the count remains the same: Mercury (0), Venus (0), Earth (1), Mars (2), and there has been no change in these numbers for decades. In 1998, Jupiter had 15 recognized satellites. The most recent at that time was Thebe, discovered in 1980 by the Voyager 1 spacecraft. In 2000, astronomer Scott Shepard and his colleagues at the University of Hawaii reported the discovery of 11 new satellites, of which nine were in retrograde "backward" orbits around Jupiter. The sizes of these satellites were estimated between two and five miles across. The most distant of these satellites is S/2000 J2, which orbits 14.9 million miles from Jupiter, taking about 800 days for a full orbit. In 2002, an additional satellite was discovered (S/2002 J1), but by 2003, 12 more (S/2003 J1–J12) were identified between 0.5 to 2 miles in diameter. Five additional satellites were discovered in March 2003, and by April another six were added. These 24 new satellites were also discovered by the University of Hawaii team and bring the total number to 58 satellites for this planet, and with the new search techniques we can predict a total of 100 satellites for Jupiter in the next few years. Meanwhile, the 11 satellites discovered in 2000 have been given formal names by the International Astronomical Union: Callirrhoe, Themisto, Megaclite, Taygete, Chaldene, Harpalyke, Kalyke, Iocaste, Erinome, Isonoe, and Praxidike, continuing a naming convention related to the mythology of Zeus (Jupiter). Saturn's 30th satellite, along with nine other satellites, was discovered in 2000 by a team of international observers led by Brett Gladman of France using the 3.6-meter Canada-France-Hawaii Telescope in Hawaii. Uranus had 15 satellites in 1998, with most of them discovered by Voyager 2 in 1986, but this number increased in 1997 to 17 satellites and then to 21 by 2003 through the efforts of a team of 11 astronomers led by Matthew Holman at the Harvard-Smithsonian Center for Astrophysics. Neptune had eight satellites including distant Proteus (72.5 million miles from Jupiter) discovered by Voyager 2 in 1989, and three more were detected in 2003 by the same team of astronomers who had detected Uranus's new satellites. Pluto retains its status as the double planet with its only moon Charon. Will more satellites be discovered as time goes on? You can count on it. Don't be surprised if textbook tables have to be redone every few years to keep up with the changes as we start detecting objects a few miles across wandering the empty suburbs of the outer planets. The current outer planet count as of March 2003 stood at Jupiter (58), Saturn (30), Uranus (21), and Neptune (17). The whole idea of naming planetary satellites is on the verge of breaking down completely, so we might have to switch to Roman numerals or some other system.

144 Where do the planetary moons come from?

Apparently, they can either get captured into orbit or formed out of the accreting gas that went into the planet itself. In the case of the larger moons of Jupiter and Saturn, the inner moon families were probably formed from material accreted by the planet into a disk. The smaller outer moons, and for instance the two moons of Mars, look like captured asteroids. Our own Moon may be material ejected from Earth and mixed with the material from a large body that collided with Earth when it was first formed. This material later accumulated to become the Moon. So, moons can have a variety of origins. But the truth of the matter is that, for most of the satellites of the outer planets, and Mars, we do not understand the physics that can cause one body to be captured by another. Basic conservation of energy and momentum laws forbid this kind of two-body capture event unless some third party is also present. Some of the moons of Saturn may actually be pieces of pre-existing moons that were shattered by a large passing asteroid. Others may have assumed their present orbits when the main planet had an extensive gaseous outer atmosphere soon after formation, allowing frictional forces to permit capture. The detailed processes involved remain a mystery to astronomers, and given all the possibilities we may never know for sure.

145 Which objects in the solar system have a magnetic field?

It's a very motley list that includes both planets and some of their satellites. The membership in this very exclusive club currently includes Mercury, Earth, Moon, Mars, Jupiter, Saturn, Uranus, Neptune, Io, Europa, Ganymede, and Callisto. The fields of these planets and satellites have been measured over the years by spacecraft such as Mariner, Galileo, Voyager, and most recently the Mars Global Surveyor (MGS). I have eliminated Venus from this list not only because it doesn't have a "bar magnet" main field but also because its crustal temperatures are so high that any fossil crustal field would have long ago evaporated. The Galileo spacecraft discovered that Io has an iron core that takes up half of the volume of Io's interior. The magnetic fields of the other Galilean satellites seem to be caused by subsurface oceans of salty water. For Mars, its field consists of surface "fossil" traces that are present in many locations where ancient lava probably flowed. Traces of this relic field were detected by MGS satellite in 2001. The gas giant planets all seem to have powerful magnetic fields generated, as for Earth, by a dynamo process deep within each planet. Surprisingly, our moon has a very weak magnetic field, about 1 percent of Earth's. This field was directly measured by Apollo 15 astronauts on the surface. It is a very lumpy crustal field that follows many mass concentrations in the crust in much the same way that Earth's crustal field does. This means that the Moon has a core about 280 miles across that may be molten at a depth of about 600 miles, but not iron-rich.

146   Why are the magnetic and rotation axis of planets never lined up?

We have no idea right now. This "principle" of misalignment seems to be a universal feature of the bodies we know about that have magnetic fields. Even pulsars and the black holes in the cores of distant galaxies play this game too. In our solar system, the misalignments span a wide range of possibilities: Mercury (0 degrees), Earth (12 degrees), Jupiter (10 degrees), Saturn (0 degrees), Uranus (59 degrees), Neptune (47 degrees). On a larger scale, the Sun has a field almost perfectly aligned with its rotation axis, but pulsars (rotating dense neutron stars) have very large tilts between their rotation and magnetic axis. The planetary magnetic fields are produced by a dynamo-like process operating within their partially liquid, and conducting, cores. What isn't understood at all is why these internal currents don't circulate around the rotation axis of the planet as all the other planetary currents do. These magnetic tilts are trying to tell astronomers something about the dynamo process, but no one seems to have a good idea what nature is trying to tell us and how to fit this important observation into our theoretical models.

147   Will other planets ever collide with Earth?

The positions of the planets in our solar system have been predicted for the next billion years. There are no signs they will ever stray more than a few million miles from their current elliptical orbits. We cannot forecast, however, whether some passing star might stumble by 200 million years from now and upset the current planetary orbits. We do not know of any stars that are within 100 light-years or so of the Sun on a collision course right now, but interstellar billiards is a very complicated game when we do not know enough about our neighborhood in the Milky Way. The composition of the solar neighborhood changes drastically every 250 million years as the Sun orbits the center of the Milky Way, and a new cast of currently undetectable stellar neighbors slide into view.

148 Can a planet ever explode?

There is no known physical phenomenon that would allow a planet, by itself, to suddenly explode, or shatter through some internal mechanism. Some low-grade science fiction stories (like the ones I often enjoy) may have used this to generate a storyline, but the truth of the matter is that no chemical or nuclear process can produce enough energy to disassemble a planet like Earth against its own gravitational binding energy. Although you can't get a fusion reaction to start in Earth's core, you can get uranium fission reactions going provided there isn't enough material to absorb the neutrons. To destroy Earth you would need to assemble a ball of uranium at the core with a diameter of about three miles. It is impossible for natural processes to produce such a concentrated object of pure uranium, and even in the assembly, so much of the interior contains neutron-absorbing elements that the reaction in all likelihood would be quenched before it started. (See Question 154.)

149 Are astronomers searching for the tenth planet?

Astronomers are always on the lookout for new members of the solar system. Not all astronomers, mind you, just those who are interested in the outer solar system. The most active programs involve searches for new comets, asteroids, and satellites of the outer planets. Part of the difficulty is in deciding what defines a planet. If Pluto is a planet, how about something two or three times smaller, making it just a little bigger than the asteroid Ceres? Varuna and AW197 are about 550 miles across, and Quaoar is 780 miles across, making them as much as 1/8 the size of Pluto. Quaoar is about 42 AU from the Sun, and two AU farther out than Pluto, which works out to be about 180 million miles. So far, there are over 600 known objects with sizes near 60 miles in diameter that have been identified beyond the orbit of Neptune. Most of these have been discovered since detailed searches began in 1992. All of them seem to be very dark bodies, probably covered with a crust that is rich in organic compounds. Astronomers can pretty much rule out an object as big as Jupiter orbiting a little further out than Pluto, since its gravitational tugs on Pluto's orbit would be pretty obvious. But a planet as large as Earth at, for instance, twice Pluto's distance from the Sun would be almost undetectable. So, the answer is that we simply don't know if a bona-fide tenth planet exists beyond the orbit of Pluto. There is so little sunlight out there that finding one would be a major technological accomplishment especially if its surface is as dark as the objects we are seeing such as Varuna and Quaoar.

150 How often do planetary alignments occur?

The May 5, 2000, conjunction was probably the most intriguing and greatly anticipated planetary conjunction in recent memory. Other kinds of planetary conjunctions happen at complicated intervals when more than two planets are involved. The last time the five planets plus Sun and Moon were as close as the May 5, 2000, conjunction was on February 5, 1962, with the planets within about 15 degrees of the sky, and there was a solar eclipse at the same time. The next time we will get a multi-planet grouping will be on September 8, 2040, when the grouping will cover an even smaller area less than eight degrees. And what about the past? On July 25, 1624, there was an "alignment" of Jupiter, Saturn, Mars, Sun, Mercury, and Venus where the distance from the Sun to Jupiter in the evening sky was only 22 degrees, and from Venus to the Sun in the morning sky, the distance was 13 degrees, with the other planets in between. This alignment was almost as tight as the May 5, 2000, event. You get about one or two of these each century. The most spectacular Grand Conjunction seems to have been on September 10, 1984, where the four planets in the evening sky were lined up within 16 degrees of the Sun! Saturn and Jupiter are also in conjunction with each other with a four-degree separation. There was a more recent Grand Conjunction on December 20, 1984, with a somewhat wider separation of about 40 degrees between Mercury and Mars. In 1981 we had a Grand Conjunction on September 10 nearly as good as the May 5, 2000, event. Its only misfortune was to not have occurred during a numerically memorable year like 2000 A.D. In 2065, Venus will transit Jupiter's disk as viewed from Earth. This will be very spectacular. In the far future, there will be some very unusual events that our ancestors will be treated to. For example, in 2518, Venus will pass in front of Saturn. The sizes of these two planets will appear about the same as viewed from Earth. In 2669, Saturn will transit the Sun as viewed from Uranus. By that time, we may actually have research stations way out there to marvel at this cosmic event. What those folks might be doing out there will be a complete mystery to us today. (See Question 16.)

151 Is there any evidence for chaotic motion in the solar system?

There seems to be a lot of it depending on where you look, and for how long. In 1987, Gerald Sussman and Jack Wisdom at MIT built a special-purpose computer called the Digital Orrery to calculate the orbit of Pluto for the next 845 million years at 40-day intervals. This is enough time for Pluto to complete 3 million orbits of the Sun. What the astronomers found was that not knowing where Pluto is today to a precision of less than one mile adds up over thousands of orbits to make the predictions change drastically. In 1989, Jacques Laskar at the Bureau des Longitudes in Paris used a sophisticated computer model to follow the evolution of the entire solar system over 200 million years. He was able to show that the orbits of the inner planets are basically chaotic as well. Chaotic behavior seems to be common throughout the entire solar system. An error by as little as 50 feet in the position of Earth today would render a forecast for its position meaningless in as little as 100 million years into the future. The orbits of Mercury, Venus, Earth, and Mars are not as stable as they seem over the long term—say a few billion years. Depending on the exact initial conditions, the cumulative influences of Jupiter and Saturn eventually cause the inner planets to have more elliptical orbits, and in some simulations, Earth, Mars, or Venus are actually ejected from the solar system. Did chaos clobber the dinosaurs? In June 2001, Bruce Runnegar, the director of NASA's Astrobiology Institute, and his team made the controversial suggestion that, 65 million years ago, a significant change in the inner planetary orbits may have allowed a large asteroid to collide with Earth. A small change in the inner planet orbits could have affected the entire solar system and made conditions favorable for a large asteroid to sneak in and decimate the dinosaurs.

152 What causes massive planets to form instead of brown dwarfs?

Planets in the solar system are believed to have formed out of a primeval disk of dust around the newborn Sun. Brown dwarfs, like full-fledged stars, would have fragmented and gravitationally collapsed directly out of a large cloud of hydrogen but were not massive enough to sustain fusion reactions at their cores. The 30 or so known brown dwarfs now include LHS 2397b, Teide 1, and Kelu-1, which are known to be cooler than about 1,300 K, and with masses the size of 50 to 75 Jupiters.

153 How far does the solar system extend around the Sun?

The Kuiper Belt in the most distant regions of our solar system is a disk-shaped system of cometary nuclei and large bodies located past the orbit of Neptune, extending perhaps 15 billion miles from the Sun (see Plate 8). It is now considered to be the source of the short-period comets. Astronomers have detected over 600 objects orbiting in this region, typically small bodies 20 to 200 miles in diameter but a few noteworthy objects such as Quaoar and Varuna have diameters of 50 and 800 miles. Table 3 is a list of some of the most distant bodies discovered as of March 2003. Further out than this is the heliopause boundary located about 85 to 100 AU (8 to 9 billion miles) from the Sun. This is where the solar wind streaming out from the Sun collides with the interstellar gas and produces a shock front like a boat plowing through water. Voyager plasma-wave instruments detected the first signs of this region in August 1992. Astronomers expect that the Voyager spacecraft will survive long enough to actually arrive at this region by the year 2004. On an even vaster scale, far larger than the domain of the Kuiper Belt or the heliosphere, is the massive and enigmatic Oort Cloud—a collection of comets-in-waiting orbiting 1,000 times farther out than Pluto. It is believed to extend approximately 18 trillion miles from the Sun, about two-thirds of the way toward the nearest star. This vast distance is considered the edge of the Sun's orb of gravitational influence. Within the cloud, comets are weakly bound to the Sun. Passing stars and other forces from the stars in the rest of the Milky Way can readily change their orbits, sending these comets-in-waiting into the inner solar system, or out to interstellar space. The total mass of the billions of comets in the Oort cloud is estimated to be 40 times that of Earth.

154 Did the asteroid belt come from a planet that blew up?

The idea of an exploding planet is the subject of bad science fiction (which I enjoy greatly) and not solar system reality. What we do know is that much of the asteroidal material was probably in a handful of larger bodies a few thousand miles across or less. These were large enough that some internal melting could happen to help segregate iron-nickel and stony materials into these two common meteorite families. Collisions between these objects eventually ground them down into the asteroids we observe today. But there could be other ways for forming the various chemical families of meteorites we know about. This subject is a very exciting, and old, research area in planetary astronomy. (See Question 148.)

155 What makes gas and rock clump together to make planets?

We can examine comet "dust" and interplanetary dust, as well as meteoritic material containing features called chondrules, and see at these scales a very important clue to their formation. The smallest materials are very delicate and in most instances are barely held together by their own inter-atomic cohesion. Carbon-rich molecules tend to make a substance very sticky, so this means that the initial growth of materials was caused by random but frequent collisions of atoms/molecules, which were chemically sticky. Eventually you end up with millimeter-sized chondrules the same way that hailstones form in thunderclouds. Dust grains develop large mantles of other compounds, especially water ice, and a variety of organic compounds that would make them even stickier during collisions. There is no physical reason why these objects wouldn't grow to sizes of several centimeters or more during frequent collisions and absorbing impacts. Once a grain nodule gets this large, the way it aerodynamically and gravitationally interacts with the gas- and dust-rich solar system accretion disk changes rather abruptly. These nodules will drift toward the mid-plane of the solar system under gravitational forces and tend to congregate there. The next step seems to be the formation of the 100-yard and mile-sized bodies that are so common in the asteroid belt and outer solar system. Physicists believe that gravitational clumping causes these particles to herd together into objects about this large. It's a process that involves not the bodies colliding with themselves but rather their acting collectively perhaps in spiral arms or Saturn-like rings. (Some asteroids we have studied with the Galileo and NEAR spacecraft seem to be little more than flying gravel banks with very little internal cohesion.) At this point, further growth depends very critically on how turbulent the solar nebula is. If it's too turbulent, these bodies will collide at average speeds that would shatter them back into nodules and dust. If the turbulence is low enough, the objects will grow rapidly by mutual collisions, and planets will form. Most of the asteroids we have seen show craters of many different sizes, so these objects also grew by absorbing smaller bodies, all the way down to chondrules and dust.

156 Have any new planets been discovered in our solar system?

The two largest bodies beyond Pluto are called Quaoar (780 miles in diameter) and Varuna (550 miles). Are these, then, Planets 10 and 11? No, because of the currently accepted convention that a planet must be at least as big as the ones we already know about. Pluto being the smallest (1,413 miles), sets the cosmic low-bar for this solar system at least. I am actually thankful that Quaoar and Varuna are not officially called planets because I find these names very hard to remember and, in Quaoar's case, almost impossible to pronounce (it is actually pronounced "kway-OH-ar"), as I am sure will millions of schoolchildren. If they are eventually called planets, and retain their current names, then we will have to invent new phrases to remember the ordering of planets in the solar system. How about "Mary's Violet Eyes Made John Sit Up Nights Pondering Quirky Verses"?

157 How can I convince children in third grade that the Sun is the center of the solar system?

There is no simple observation that leads to this conclusion if you try to follow a strict historically accurate line of reasoning. If there had been, we wouldn't have spent 5,000 years before Copernicus debating it. Let's take advantage of what we know today to give a simple and compelling proof that should satisfy any child. We know the distance to the Sun, independently of heliocentric or geocentric beliefs. We can use this piece of information to determine the size and therefore the mass of the Sun. It's huge! It is over one million times the mass of Earth. This means that, to avoid the absurdity of a puny Earth moving a massive star, we have to accept that the Sun is what is controlling the whole shebang. This leads to the sensible result that the Sun and not the Earth is at the center of the solar system. I think this is a simple idea that a third-grade student can understand.

158 How many stars produced the elements in our solar system?

So far as I know, only one star has been singled out. About 100 million years before the solar nebula collapsed, a nearby supernova deposited radioactive aluminum-26 atoms into it. This isotope decays in only a few million years, and its traces can only now be found in the enhanced abundances of its more stable "daughter" elements. This single event probably did not deposit *all* of the heavy elements into the solar nebulae, because we think that the element enrichment of the interstellar medium is a long process and requires many supernovas to do the job. However, when we look at distant young galaxies, we see that their element abundances are almost those of our Sun, and that this happened within a few billion years after the Big Bang. It only takes ten million years to form a massive star and to have it detonate as a supernova, so in one billion years, you can have many supernova events pollute any given interstellar cloud. In our case, we do not know what elements it started with or how many supernovas it took to cook them! Studies of interstellar dust grains embedded in meteorites have, however, shown that the silicate dust grains thought to be produced by ancient red giant stars come in at least five distinct isotopic families that could mean that at least five different ancient stars were involved in forming the dust grains in our solar system, and the silicate compounds so common on our planet's surface. Evidently, it really does "take a village" to raise a child—or in our Sun's case, create one. (See Question 214.)

159 Is our solar system typical of other planetary systems we know about?

Recent evidence shows that our solar system is pretty unusual. In 1997, when I first answered this question, I took the party line that there should be some basic properties of our solar system that ought to be common to others. The most distant planets in every solar system ought to be gaseous and icy bodies because ices and gases will be the most common compounds present under low temperatures in these distant regions. Now that we have detected planets orbiting over 90 stars we see that gas giants in elliptical orbits are common in inner solar system regions as well, and circular orbits are rather uncommon. As for the number of planets and their masses, that depends on the details of when and where they were formed. Some solar systems might only have a single massive planet, on the verge of being a binary star system; others might have only asteroidal rubble.

160 On which planets would it be possible to live?

Practically none of them. Jupiter, Saturn, Uranus, and Neptune have no solid surfaces, so that eliminates them as potential landing sites. Venus has such a thick atmosphere it would crush you like an eggshell. The rest of the bodies, such as Mercury, Mars, Pluto, and the satellites of the giant planets, are possibilities, but you would have to provide your own atmosphere to breathe and probably live below the surface to protect yourself from the harmful radiation and cosmic rays from the Sun. The surfaces of the icy moons would be a challenge. Humans require heat, and this would eventually cause any colony to slowly sink into the icy crust if not properly insulated from the ground. By far, the least expensive planet to visit is Mars.

CHAPTER 5

The Stars—
Celestial
Cities of Light

We have known about the stars for longer than we have recorded our history—how could we not? Not one of us has failed to be struck by the mystery of these celestial lights. They are the stuff of enchanting legends and heroic sagas writ in the scattered lines and dots of Heaven. A century's work by thousands of seekers has now reduced the stars to simple matter. We mark their diversity and passing lives in relentless tables and formulas. We learn their names and their peculiarities, as we search among them for the shadows of planets like our own. If our solar system is the eventual playground of humanity in space, the stars are the remotest outposts of our dreams of exploration. No human technology, or political will, can withstand the century-long journeys that we will have to marshal to send either ourselves or our robotic technology on such far-flung voyages. But we can dream.

● ● ◑

161   What planet outside our solar system do we have the most information about?

For now, the hands-down favorite would have to be HD 209458b. We know its mass (0.7 Jupiter's), its diameter (1.3 times Jupiter's), its density (0.27 gm/cc), its orbit period (3.5 days), and its distance from its star (4 million miles), and from all this we can determine its surface temperature (2,000 F). Thanks to the tilt of its orbit plane, this planet eclipses its star HD209458. In November 2001, astronomers detected sodium in its atmosphere. Then in March 2003, an international team of astronomers led by Alfred Vidal-Madjar of the Institut d'Astrophysique de Paris announced that this was a planet in trouble. Additional transits across its star observed by the Hubble Space Telescope had allowed them to extract even more information about this body. The planet is so hot and close to its star that its outer atmosphere, rich in hydrogen, is boiling off the planet and probably forms a comet-like tail behind the planet opposite the star. The hydrogen atmosphere of this planet extends over 124,000 miles downwind of the planet. The Vidal-Madjar team estimates that the amount of hydrogen gas escaping HD 209458b is at least 10,000 tons per second, but possibly much more.

162 How would we detect the atmosphere in a planet orbiting another star?

What a difference a few years have made! We can now speak from actual experience, having accomplished just this near-miraculous feat! In one instance, HD 209458b, the planet passed across the face of its star, and astronomers were able to detect the absorption of sodium in its atmosphere. On November 7, 1999, Greg Henry at Tennessee State University observed a 1.7 percent dip in the star's brightness. By using the Hubble Space Telescope STIS spectrometer and the forecasted transit times for this planet, Timothy Brown at the National Center for Atmospheric Research (NCAR) and David Charbonneau of the California Institute of Technology detected the traces of the element sodium—an achievement that was announced at a NASA news conference on November 27, 2001. An artist's rendering of this event is shown in Plate 9. With the spectacular success of this method, NASA will be launching the Kepler spacecraft in 2006. Its mission is to monitor the brightness of 100,000 stars. We expect it will discover as many as 900 Earth-like planets if the incidence of orbits favorable for transits is similar to what we have currently seen in the actual data. Kepler will be able to test the hypothesis that stars have terrestrial planets within the habitable zone. It will lay the groundwork for a subsequent NASA mission called the Terrestrial Planet Finder, to be launched sometime after the final selection of its design in 2006. Its mission is to image the planets discovered by Kepler and determine the atmospheric compositions for many of them.

163 What methods are used to detect planets around other stars?

Since the mid-1990's, a number of very different techniques all seem to be very good at finding planets, although many of them do not study the same planets around the same stars. *Pulsar Timing:* Pulsars are very precise clocks in the universe, and by studying them you can detect planet-sized objects orbiting them by the way the pulses arrive in time. *Transit photometry:* You can watch for the very slight diminution in light from the star caused by the planet passing in front of it. *Spectroscopy:* You can measure the slight changes in the star's velocity as the planet orbits the star over weeks to years. *Gravitational Lensing:* You can monitor the brightness changes of a star and see if they match the changes caused by the gravitational field of a planet passing between you and the star. So far, all of these methods have been used to discover over 100 planets. Transit photometry and gravitational lensing are not limited by distance so these methods, when you are lucky, can detect planets around stars thousands of light-years away. The spectroscopic method requires a lot of light from the star and is limited to the brightest and nearest stars to the Sun. NASA will launch the Kepler mission in 2006, which will watch for transits occurring in 100,000 stars, and will probably detect hundreds of new planets in a few years' time. Follow-up spectroscopic studies will then study the atmospheres of the transiting planets.

164   Have planets in binary star systems been detected yet?

Yes. Gamma Cephei is a star in a binary system where the stars take 40 years to orbit each other. In 1992, astronomer Gordon Walker at the University of British Columbia reported short-term velocity changes in this star with a period of 2.52 years but eventually dismissed this as an artifact of the star's rotation. The changes would have been consistent with a Jupiter-sized planet orbiting about two AU from the primary star. A 20-year follow-on study to this variability by astronomer William Cochran at the University of Texas led to the announcement in October 2002 of a body with a mass of 1.25 Jupiters located 1.8 AU from Gamma Cephi. Since the distance between the primary and the secondary stars is 12 AU at their closest, this is the first planet detected in a compact binary system where the planet doesn't instead orbit both stars at a far distance. The primary star is a K1 dwarf with a mass of 1.5 Suns, which would look orange in color. The cooler companion star with one-fifth the Sun's mass orbits at a distance of about the planet Saturn and is a red dwarf star. It is a good question how such a planet could have formed and persisted in this system for several billion years with one star near where our sun is and a second out where our planet Jupiter roams. It doesn't leave much room in between.

165   How can we possibly know how old a star really is?

For decades we have done this by understanding the details of how stars produce their energy by thermonuclear fusion, and by using powerful stellar evolution calculations that match the observed properties of a star to nuclear physics timelines. In recent years, astronomers have even accomplished the unthinkable. In February 2001, a group of French astronomers led by Roger Cayrel François from the Observatoire de Paris announced the first dating of the age of a star using radioactive techniques. The group used the ESO Very Large Telescope in Paranal, Chile, to analyze the amount of uranium-238 in one of the oldest stars in the Milky Way: a 12th-magnitude star known only by its catalog designation CS 31082-001. Based on this uranium-dating technique, the astronomers determined that the star is 12.5 billion years old. This age was in the ballpark previously estimated for this star based on the low amount of other "heavy elements" such as iron. It has 200 times less of these elements than the Sun and would have made a poor candidate for forming planetary bodies. Fourteen lines of thorium-232 and one strong line of uranium-238 were detected, and from these it was possible to radioactively date this star in much the same way that archeologists use carbon-14 and carbon-12 to date ancient human artifacts and remains.

166 What is the largest star that astronomers know about?

When we say "large" for a star, it is a rather ambiguous term. With rare exceptions, the largest stars in terms of diameter are among the very oldest M-type red supergiant stars that are near the end of their lives prior to the supernova phase. The largest stars by mass are all among the hottest O-types and have lives measuring only a few million years. Table 4 is a list of some examples of large stars based on their physical size and their masses. VV Cephi and Mu Cephi have the biggest known diameters while the Pistol Star in the Galactic Center seems to be the most massive and tips the scale at over 200 times the Sun's mass. Mu Cephi, also known as Herschel's Garnet Star, is a red supergiant star about 1,800 light-years from Earth and spanning a size of nearly 2,500 times the diameter of the Sun. That would place its cool outer layers just beyond the orbit of Saturn. The Pistol Star shines with ten million times the power of the Sun and would fill Earth's entire orbit with its mass. It has a powerful wind that constantly hurls gas from its surface in such volumes that every 10,000 years it loses an equivalent mass equal to the Sun's. It is destined for a spectacular supernova demise in perhaps 1 to 3 million years from now . . . or sooner. It was discovered by astronomers using the Hubble Space Telescope's NICMOS infrared camera in 1997, but had been spotted as early as 1990.

167 What is the latest count on the number of extra-solar planets?

According to tabulations by Jean Schneider at the excellent web site "Exoplanet Encyclopedia" (http://www.obspm.fr/encycl/encycl.html), the total as of June 17, 2003, was 108 extra-solar planets, located in orbit around 94 normal stars and two pulsars (PSR 1257+12 and PSR B1620-26). There are now 12 multiple-planet systems containing at least two planets. The list includes one planet orbiting a G-type star, detected via the new technique of gravitational lensing (OGLE-TR-56 b) at a distance of 4,900 light years from the Sun. The planets around the normal stars range in masses from about one-fourth that of Jupiter to 14 times its mass, with most of them in the range from one to three times as massive as Jupiter. These mass estimates are only upper limits because we do not as yet know the inclination of the planetary orbits with respect to our line-of-sight. In two cases (HD 209458b and OGLE-TR-56b), we do have this information because the planets passed in front of their stars as viewed from Earth. The masses are 1.4 and 0.9 times Jupiter. The closest "exoplanets" are located in orbit around the stars Gleise 876 (13.7 light years) and Epsilon Eridani (10.4 light years). Epsilon Eridani is a variable K2 dwarf star with a mass of 0.8 times the Sun's. It has one definite planet (b) that has about 0.9 Jupiter masses and orbits 3.3 AU from its star in a near-Jovian orbit. A second planet is suspected in this system with a mass of 0.1 Jupiters and a distance similar to that of Pluto (40 AU), but there is some debate over this detection. Because of its distance from the star, Epsilon Eridani b must be very similar to Jupiter, and with a star whose luminosity is one-third that of our Sun, temperatures are a bit chillier for the satellites that may orbit this planet. The star Gleise 876 is a faint, 11th-magnitude M4 dwarf star with a mass only a third that of the Sun. It has two planets orbiting at 0.13 and 0.21 AU from the star, with masses of 0.56 and 1.89 Jupiters. These planets orbit closer to their star than the planet Mercury orbits our own Sun (0.35 AU), but because their star has a luminosity of only 1/80th of our Sun, their atmospheres must be at temperatures well below those of Mercury (840 F) and very likely are in a range where liquid water may exist if present.

168 When will astronomers start to name all those planets they are finding around other stars?

Since 1998 when astronomers began to find extra-solar planets in large numbers, the naming of these new worlds has followed an informal convention set up in a hurry to handle all the new discoveries. It is based on a scheme that makes a lot of sense, but it does lack some of the beauty of the other naming conventions we use for bodies within our solar system. If a star is discovered to be a part of a binary star system, the main star receives an A designation and the companion stars receive B, C, D, etc. For example, Alpha

Centauri is the main star of a three-star system. It is given the name Alpha Centauri A and its companions are called Alpha Centauri B and Alpha Centauri C. Objects such as planets that are discovered around these objects will be named with lower-case letters, and this is where it gets interesting, because these names follow the timeline of historical discovery, not distance from the star. Suppose from a distant star, our Sun was called Alpha Zymergy. Then in order of the ease of detection, our planets would be called Alpha Zymergy a for Jupiter, Alpha Zymergy b for Saturn, Alpha Zymergy c for Uranus, and Earth would be Alpha Zymergy f. This naming scheme is not going to endear itself to science fiction readers. In fact I find it very clunky and unpleasant, but I respect the logic of it just the same. Will we eventually rename all these new worlds according to a more poetic and evocative scheme? Probably something like this may happen. Astronomer Alan Boss, who is the chairman of the International Astronomical Union's Working Group on Extrasolar Planets, thinks that some decision will have to be made very soon on a formal naming scheme. And by the way, this scheme will probably not deal with the potentially hundreds of free-floating planets found in the Orion Nebula and elsewhere. No one has a good idea of what to name them either.

169   Could the Hubble Telescope see planets around other stars?

Under certain circumstances and with the right instruments, it could. Planets can be detected by monitoring the frequencies of many spectral lines from the star to a Doppler speed precision of 30 feet per second. Planets could be detected by their reflected light by the Hubble Space Telescope, because the optics and "seeing" conditions on orbit are superior to anything on the ground looking up through the turbulent atmosphere. A Jupiter-sized planet orbiting a star three light-years away at the distance of Jupiter from the Sun will be seen about five arc seconds from the star. Only Hubble's perfect optics and no distortion will allow astronomers to search for faint objects within a few arc seconds from a nearby star. However, few searches have been attempted because other research projects have higher priority.

170   What do stars look like up close?

A lot like our Sun, but with many interesting differences and of course color changes. Since the early 1980's, astronomers have developed methods that allow them to see the surfaces of many nearby stars, and of more distant red super giant stars like Betelgeuse. The large stars have very large sunspots—as big as our Sun and larger, while smaller stars like our Sun seem to have magnetic fields threading their outer layers. The "starspots" come and go in decade-long cycles as with our Sun. Of course the temperatures for stars are different, but that seems only to affect the "color" of the star's surface. So long as the outer layers are convecting like boiling oatmeal, this activity seems to be enough to kick up the same kinds of phenomena we see on the Sun including prominences and flares. There are stars like Vega that probably don't have convection zones near the surface. These stars will probably have very bland surfaces containing a few very large convection cells, but with few if any small-scale features.

171 Is there any recent information about the Star of Bethlehem?

There have been many "theories" about this dramatic celestial event. Curiously, only one author of the New Testament mentions it at all: Matthew. Over the years, these explanations have included comets, supernovas, and planetary conjunctions. No mention of the Star of Bethlehem occurs in Chinese records, and by then the Chinese were accomplished sky observers. How do you "hide" a phenomenon like the Star of Bethlehem from the rest of the world? This immediately discounts supernovas and spectacular comets from any reasonable list of possibilities. These were usually portents of evil or doom, anyway! The idea that it was an unusual planetary conjunction seems to satisfy all known constraints and data from astronomical sources. Robert Molnar at Rutgers University has recently given the many planetary conjunctions that have been offered as matches between 6 B.C. and 1 B.C. a completely new analysis. He had been studying the astrological importance of Aries the Ram in the ancient world and originally had no intention of even addressing the Star of Bethlehem issue. His research, however, soon led him to astrological texts from that era and to the fact that Aries was the zodiacal sign often representing the Jews. If there were to be a Messiah who was the King of the Jews, then, astrologically, the kingship planet, Jupiter, must be involved in this sign, too, he reasoned. In fact, on April 17, 6 B.C., Jupiter underwent the first of two occultations (eclipses) by the Moon in Aries in 6 B.C. The presence of the Moon also amplifies the kingship event. As Molnar explains, "The second occultation on April 17 coincided precisely when Jupiter was 'in the east,' a condition mentioned twice in the biblical account about the Star of Bethlehem. In August of that year Jupiter became stationary and then 'went before' through Aries where it became stationary again on December 19, 6 B.C. This is when the regal planet 'stood over'—a secondary royal portent also described in the Bible." The terms "went before" and "stood over" are astrological terms translated verbatim and refer to the retrograde motion of a planet and its stationing in the constellations. Nothing more, and nothing less. Once the Three Wise Men (who were astrologers) realized what was happening, they had more than enough time to mount a caravan and travel to Bethlehem over the course of several months' time. Jesus was born in the spring, when sheep were having their kids and shepherds were watching through the night. This would have happened around April 17, 6 B.C.

172   If stars come in colors, why do most appear white to the eye?

When you look up at the sky on any clear night, just about all the stars look white. Only a very few of the brightest ones appear red, yellow, or orange. There is a simple reason for this. The eye has two kinds of light receptor cells: rods and cones. Color vision comes from the response of the cones, which, however, are not very light sensitive. Below a fixed light intensity, they stop responding at all. The rods, on the other hand, are very sensitive photoreceptors even at low light levels, but they are monochromatic, sensing only black and white. When you look at the brightest stars with the naked eye, or when you use a telescope to concentrate the light that falls on your retina, the light can be above the threshold at which the cones can register color, but for the thousands of other stars in the skies fainter than these brightest stars, only the monochromatic rods see anything and do not pass on color information to the visual cortex of the brain.

173   What would happen if two stars collided?

There are several possibilities. If the collision speed is higher than a particular threshold speed, say about 300 miles per second, enough kinetic energy would be imparted to the two masses that the stellar material would dissipate into a vast expanding cloud of gas, never to reassemble itself into a new star. If the speed were very slow, the stars would merge into a new, more massive, star. The evolution of the new star would begin with a rejuvenated core of fresh fuel because the merging of the two stars would have mixed new hydrogen fuel into the core of the new star. If the speed of the impact is moderate and off center, the stars will go into a very tight orbit around one another, perhaps even sharing a common gaseous envelope. Over time, the two separate cores would spiral into each other, and you would again be left with one new, massive star. Since the escape velocity of the Sun is about 1.3 million miles per hour, this is about equal to the threshold speed of the impact. If a smaller star, like a white dwarf or neutron star, smashes into a bigger star, like a red giant, most of the giant's outer envelope would be blown off as it absorbs the impact. The results get a little more violent when two smaller stars collide. Neutron stars are very small and dense. If a neutron star reaches a certain mass, it will implode and form a black hole. Therefore, if two neutron stars merge but their combined mass is more than the maximum mass a single neutron star can have, they implode into a black hole. If the circumstances are the same when two white dwarf stars collide, they will implode into a neutron star, or even a black hole.

174   Is there life outside the solar system?

Right now, we have absolutely no data to suggest that life exists outside our solar system. Still, astronomers are putting together a persuasive line of evidence that planets are common around other stars. In our particular solar system, there are several spots where life could have gotten a fleeting foothold, and whose traces may even exist today: Mars, Europa, Titan. So, I think it is very plausible that at least bacteria exist on planets orbiting other stars. Higher forms of life . . . well . . . that's a very hard nut to crack even if we do detect free oxygen in the atmospheres of distant planets. We had a very pleasant oxygen atmosphere on Earth for 500 million years before radio technology was invented. Astronomers have studied the 90+ stars we know about with planets, and have calculated the sizes of the habitable zones where a planet like Earth could exist in a circular orbit, even with Jupiter-sized planets orbiting in elliptical orbits. According to astronomers Barrie Jones and Nick Sleep in the United Kingdom, the star 47 Ursa Majoris presides over the most solar system–like of the handful of planetary systems they have studied mathematically. It has a rather wide habitable zone and a Jupiter-sized planet that keeps far away from this region. If there is an Earth-sized world there, it may be a lovely little garden spot to visit someday.

175   How can planets be present around pulsars, which are exploded stars?

You would think that, when a star goes supernova to produce a neutron star, a delicate thing like a planet would be blown away like dandelion seeds in a gale. That seems not to be the case in two separate instances. In 1991, after a four-year study, astronomers Alexander Wolszcan of the Pennsylvania State

University and Dale Frail of National Radio Astronomy Observatory discovered among the details of the radio signals from the pulsar PSR B1257+12 that the signal's changes could be explained if two small planets orbited the pulsar. A few years later, the traces of a third planet emerged from the data. The planets would have to have masses of up to three times that of Earth, and orbits with distances of 9 to 40 million miles. A possible Jupiter-sized body may also exist orbiting at Pluto's distance with a period of 170 years. Since then, a second pulsar has been added to the short list of neutron stars with orbiting planets. But how could the devastating supernova explosions that surely accompany the formation of neutron stars have left these planets intact? The orbits also seem to be exceedingly circular, as though not much has happened to disturb them in billions of years, yet the pulsars are probably less than 800 million years old of themselves. The origins of these planets remain very much a mystery today. If a star loses more than half its mass, a sure bet for most supernovas, any orbiting planets will drift out of the star's gravitational grasp. These orbits should be highly elliptical, but, in the case of the actual planets, they are not, so this process could not have led to the formation of the planets. The planets could not have formed before the supernova detonation occurred. At the present time, there is no accepted explanation for how these planets formed after the supernova.

176   Can you tell me a little bit about the Pleiades?

The Pleiades star cluster in the constellation Taurus is a cluster of 250 stars of which nine have been seen and named by careful observers for centuries. Their names are Alcyone, Atlas, Electra, Maia, Merope, Taygeta, Pleione, Calaeno, and Asterope. The cluster actually contains over 3,000 stars of all masses and luminosities, at a distance of 410 light-years. It is a young cluster probably no more than 20 million years old, which still has some of the vestiges of its infant dust cloud visible as the faint haze that surrounds this cluster of stars. The cluster moves across the sky at a speed equal to the diameter of the full moon every 30,000 years. The Pleiades has been mentioned in one way or another for thousands of years. As long ago as 2357 B.C., there is reference to it in ancient Chinese texts. In ancient Aztec and Mayan tradition, its midnight culmination directly overhead every 52 years was an event of great concern since it was believed that the world would come to an end on one such occasion. The Pleiades goes by a great many different names—The Seven Sisters, The Children of Atlas, or the Daughters of Pleone, according to ancient Greek mythology. It is mentioned in the Book of Job and in the works of Middle Eastern poets such as Hafiz of Persia (14th century). The Pleiades is associated with water, sailing, and certain festivals such as Halloween and various memorials to the dead. Along with the Mayans, other civilizations celebrate its midnight culmination as well. Native American legend has it that a great bear chased seven maidens into the sky near the famous Devil's Tower monument in Wyoming. The monument still bears the scars of the claw marks of the bear as it tried to get the maidens. A popular name for it in German folklore is the Sailor's Stars. Farmers in Germany refer to this cluster as the Clock Henne and Her Chickens. Other lands associate it with a flock of pigeons, camels, goats, or doves. It is the Hoening Stars in South Africa; the Abipones of Paraguay consider it to represent beloved spirits of their ancestors. In the Middle Ages, it had a more sinister influence as the Witches' Sabbath or the Black Sabbath since the Pleiades culminates overhead on November 1 at the time of the Druid celebration. A number of Greek temples are apparently oriented to the rising or setting of this cluster, and in Egypt it is revered as the Stars of Hathor, one of the many forms of Isis. One of the oldest traditions is that of the lost Pleiad. The Greeks identify her as Electra, or Merope. According to Aratus, "Their number seven, though the myths oft say, and poets feign, that one has passed away." It is a tradition also found in ancient Japanese lore, the writings of Australian aborigines, natives of the Gold Coast of Africa, and headhunters of Borneo. Pleione is the most likely candidate for the lost Pleiad since it is a variable star.

177   How far away is Proxima Centauri?

Proxima Centauri is a dwarf M5 emission-line star with a luminosity of 1/17,000 times the Sun. It was first discovered in 1913, when it was at that time 0.1 light-years closer to the Sun than Alpha Centauri. The mass of this star is 0.1 the Sun with a radius probably about half that of the Sun. According to the Hipparcos satellite measurements, Proxima Centauri is at a distance of 4.22 light years from Earth. It is a member of the Alpha Centauri star system. Alpha Centauri A and B at 4.40 light years. Proxima will remain the closest star to the Sun for about 500,000 years or so. The system moves across the sky an amount equal to the full moon diameter every 500 years. Since it is so close, have astronomers ever detected planets orbiting it? A team led by George Benedict at the McDonald Observatory in Texas used the Hubble Space Telescope in 1995 to search for signs of planets but could only report upper limits. The team collected 52 sets of observations of a star field containing Proxima Centauri using the Fine Guidance Sensor. They would have been able to detect a Jupiter-sized planet with a period between 90 to 600 days if it were present. This would place such a planet between the orbit of Mercury and Mars. No indications of such a body were found.

178 Would a nebula look colorful if you traveled into it?

No. You would probably see nothing at all because the gases are so spread out in space. At a density of only a few hundred atoms per cubic centimeter, most nebulae are better than the best vacuums we can create under laboratory conditions, and as such it would be impossible to see their color if you were inside one of them. I am always amused by movies that portray star ships inside or near very colorful nebulae, or with background skies swirling with color. *Star Trek* movies love to hide star ships in nebulae, but you would just as soon hide a mosquito in a light bulb. The typical distance between the star ships is seldom less than a few million miles, and a nebula dense enough to hide them would very likely be dense enough to collapse into a black hole.

179 Do stars move?

They sure do! Think of our neighborhood in the Milky Way as a giant roadway that encircles the galactic center. The stars are cruising at speeds of a half a million miles per hour. Like an ordinary highway filled with traffic, you sit in one of these cars and see your neighbors moving near you. Some are pulling ahead and some are falling behind as they try to maintain an average speed. Unlike our roadways here on Earth, the Sun and its neighbors can continue circulating along this cosmic beltway for billions of years without a single fender-bender. Near the Sun, the typical relative speeds can be about 10,000 miles per hour, but interlopers from the halo regions of the Milky Way can travel a dazzling 200,000 miles per hour. Our Sun is dragging our solar system in the direction of Vega at a speed of about 43,000 miles per hour. If Vega obliged us like a car in the breakdown lane and stayed put at its current distance of 26 light-years, we would catch up to it in, oh, about a million years. One of the fastest known stars in terms of its local speed compared to the Sun is the recently discovered neutron star RX J185635-3754, located 200 light-years from the Sun. It is a "runaway" from a supernova explosion over one million years ago and is traveling through space at nearly 240,000 miles per hour. It will reach its minimum distance to Earth of 170 light-years in about 300,000 years.

180   Why do stars evolve?

Although a star like the Sun looks like some vast globe of gas frozen in time, it is actually a very delicate creation of nature, reminiscent of a snowflake on a winter's day. It is a delicate and exact balance between two powerful forces in nature. Gravity pulls on every gram of a star's mass to force it downward and inward to its core. At the same time, the pressure caused by heated gas pushes relentlessly outward and wants to dissipate the star in a powerful explosion, like air escaping from a flat tire. In many ways, gravity defeats its own purpose because as gas is forced core-wards, it gives up the energy it gained from the inward fall and converts this into heat. The more gravity tries to compress the gas, the more energy the gas receives from the infall, and the hotter the core becomes. As a star is formed from collapsing matter, eventually the core reaches a point where the gas atoms collide with enough energy to actually fuse their nuclei into heavier elements such as helium. This releases more energy, and slowly the core begins to push back against gravity to slow down, and then finally stop, the in-rushing flow of gas. Not only is a star born in this way as a stable object, it now begins a constant battle between gravity trying to collapse the star and thermonuclear fusion, which produces internal pressure. As the eons pass, this balance slowly tips in favor of gravity as the core can no longer supply fuel to run the nuclear furnace. The star has no long-term option but to change its properties and evolve. This is why stars change throughout their lives, because of the changing balances between internal fusion pressure and gravity's crushing forces.

181   When we see stars, have they already burned up?

Not the ones we can see with the naked eye on a clear night. In our galaxy, virtually all of the stars we see still have millions of years to go before they "burn out." The stars we see in our own galaxy are too close for them to be greatly different than what we see, except for the massive red giant stars that are on the verge of becoming supernovas. Betelgeuse in Orion may already have gone supernova for all we know, but we won't know for another 1,500 years. Now, the Hubble Space Telescope is studying variable stars in galaxies tens of millions of light-years away. In ten million years, such stars can change a lot, and very likely some of the ones they have measured no longer exist and have become white dwarf stars already. We also see very distant supernovas in galaxies hundreds of millions of light-years away. These individual stars have already exploded, and by now have been black holes or neutron stars for millions of years.

182 How long does it take for a star to burn out?

This depends on its mass. To evolve off the hydrogen-burning main sequence takes approximately 90 to 95 percent of the lifetime of the star. After this long "middle-age" period, a star evolves very quickly and is "dead" as a supernova, a white dwarf, or other cinders, after a time equal to about 10 percent of its middle-age span. A massive O-type star like the Pistol Star in the center of the Milky Way lasts about five million years and ends up as a black hole. An A-type star like Sirius lasts about 500 million years. An F-type star, Canopus, makes it to about two billion years. A star like the Sun lasts about 12 billion years. A K-type star like Aldebaran can last 20 billion years. A high-mass M-type star like Betelgeuse can make it to about 50 million years, and a low-mass M-type star like Wolf 359 can last 100 to 200 billion years. Brown dwarf stars such as GL 229B have the lowest possible stellar masses and can last nearly 500 billion years by frugally fusing their fuels at low temperature.

183 Why do planetary nebulae have hourglass shapes?

Planetary nebulae happen when a red giant star begins to lose a substantial amount of its outer layers toward the end of its evolutionary phase and just before its core is exposed as a white dwarf. You would think that a round star should produce a round (spherical) cloud of gas as a planetary nebula, but this is usually not the case, as you can see in Plate 10. The details are not fully understood at this time. Magnetic fields are one possibility that change "smooth" to "complex." Another possibility involves the ejection of fast-moving material within a pre-existing outflow of gas from the same star or from a companion. Astronomers speculate that the interaction between these two stars may have sparked episodic outbursts of material, creating the gaseous bubbles that form the nebula. They interact by playing a celestial game of "catch": As the red giant throws off its bulk in a powerful stellar wind, the white dwarf catches some of it. As a result, an accretion disk of material forms around the white dwarf and spirals onto its hot surface. Gas continues to build up on the surface until it sparks an eruption, blowing material into space in a nova-like event. This material travels more easily in directions perpendicular to the accretion disk and is "corseted" along the plane of the disk, so you end up with what are called bipolar planetary nebulae. But there are so many varieties of planetary nebulae that one mechanism probably doesn't fit all. There is still much we don't understand about these beautiful objects.

184   Which is the coldest white dwarf known?

A white dwarf is the compressed core of what once was a red giant star. It is the leftover cinder from a star like our Sun when it has ended its life and no longer can fuse elements together to provide energy to sustain it as a normal star. White dwarfs start out life as very hot bodies but rapidly cool in a matter of a few thousand years until they resemble white-hot stars but producing a thousandth of the Sun's light power. Over time they will continue to cool until they reach the temperature of space itself (2.7 K), but this will take hundreds of billions of years. So, the coolest white dwarf stars we can find today are also the oldest white dwarfs and were created as Sun-like stars died billions of years ago. The coolest known white dwarf is WD0346+246 with a temperature of only 5,800 F. It was discovered in 2000 in the constellation Taurus by Dr. Nigel Hambly at the University of Edinburgh along with a team of six other astronomers. It is located 90 light-years from the Sun and has a probable age between 7 and 11 billion years. Since stars began to form in the universe about 13.5 billion years ago, the progenitor star must have resembled Sirius, which will become a white dwarf in a billion years or so.

185   How fast would a neutron star have to spin to break up?

A 24-mile-diameter neutron star equal in mass to the Sun would need to rotate at 27 percent the speed of light for the gravitational force to equal the centrifugal force at its equator. This means 1,000 rotations every second. This is only slightly faster than some of the known millisecond pulsars that are identified as being very young. Pulsars seem to be born as very fast-spinning objects immediately after a star explodes as a supernova; then they slow down very rapidly as they age. After a few thousand years the spin rate is only about 30 times a second. They may slow down by a process called magnetic breaking. The powerful neutron star magnetic field connects the pulsar with the surrounding supernova remnant gases, and rotational energy is transferred into these gases. Another possibility gaining a lot of attention, thanks to general relativity calculations and computer modeling, is that the fast-spinning neutron star generates huge amounts of gravitational radiation before it settles down to its perfectly spherical shape. This also slows the spinning down over time.

186   Do all O-type stars become supernovas?

O-type stars are universally more than 20 times the mass of the Sun, and virtually all evolutionary models predict that they will become what astronomers call "Type II" supernovas within about 100 million years or so, depending on their masses. Even B-type stars such as Rigel with masses from 6 to 20 solar masses probably do not escape this fate but do not leave behind black holes. They seem to be the factories that create neutron stars.

187 What is a neutrino star?

Not to be confused with neutron stars, a neutrino star is what some very massive stars probably become when they are just about to detonate as a supernova. The energy emitted by such a star into interstellar space is not dominated by light radiation, but by the enormous energy carried by neutrinos produced in the hellish, billion-degree cores of these stars. The neutrinos carry off so much energy that the core fusion reactions eventually cannot keep up with this vast energy drain. The process of collapse is triggered, and this causes a supernova.

188 What would happen if Betelgeuse went supernova?

Betelgeuse is about 500 light-years distant. If we just consider what could happen as a result of its expanding shell of gas, the picture is quite dramatic. Typical shell velocities are about 20 million miles per hour. The shell traveling slower than the speed of light would arrive here about 100,000 years after we see the star explode. The shell would carry perhaps ten times the mass of the Sun. The flow of particles at our distance would be about 20,000 protons per second per square inch. The solar wind flow, by comparison, is about 50 million protons per second per square inch at Earth's orbit. So, although detectable, the flux from Betelgeuse probably won't do much biological damage compared to what the Sun does. However, because the Betelgeuse flow is traveling at 20 million miles per hour compared to the one million miles per hour of the solar wind, the Betelgeuse flow has an effective pressure that is thousands of times stronger than the solar wind and spread out over a region much larger than the size of the entire solar system. This would cause the Sun's heliopause to collapse from its present radius near 100 AU, to possibly less than the orbit of Earth. Also, Earth's magnetosphere would be compressed. If the Betelgeuse flow were applied directly to Earth's magnetosphere, it would be pushed into Earth's atmosphere, leaving all of our satellites exposed. The energies of the particles in the Van Allen belts would be amplified during this compression. The environment outside the Van Allen belts would probably be "lethal" for human exploration of the solar system. There are also

X rays to contend with. Inside the supernova shell, we would be subjected to this high-energy radiation for tens of thousands of years. These X rays would not get down to Earth's surface thanks to our thick atmosphere. Travelers in interplanetary space would need some additional shielding as these X rays strike the skin of their spacecraft and liberate fast-moving electrons that penetrate into the living quarters and spacecraft electronics.

189   Are supernovas the same as novas, but just brighter?

They are quite different, and typically differ by a factor of a million in their power. Novas can be either recurrent or non-recurrent. That means they can either reappear many times or be a one-shot event for that star. About 40 novas per year occur in the Milky Way compared to one supernova every 40 years. Novas seem to be binary stars in which one of the companions is a dense stellar remnant like a white dwarf. The other member of the system is an evolved red giant star millions of miles across, perhaps even bigger than Earth's orbit. Some of the gas from the evolved star makes its way over to the dense companion, forming an orbiting accretion disk. Gas from this disk then leaks onto the surface of the companion and builds up until it reaches some critical mass. The surface of the dense companion then explodes in thermonuclear violence as this fresh new gas is consumed and ejected to produce an expanding cloud. The gas disk can be disrupted by this explosion, but over the years it rebuilds and the process repeats itself as a recurrent nova. There are also one-shot novas that do not seem to recur.

Supernovas come in two major flavors: Type I and Type II. The Type I supernovas are commonly found in elliptical galaxies and are believed to have as their precursors a nova-type binary system. They are associated with stars that make up the oldest low-mass stars in systems such as globular clusters. Evidently what has happened here is that the star has evolved into a red giant, and some of its gas was absorbed by a dense companion star, probably a white dwarf or even a neutron star. But instead of a gentle nova-like outburst, the companion star was completely blown apart by the detonation. Type II supernovas come from the most massive stars in a galaxy that are at least eight to ten times more massive than the Sun. In this case, the star evolves until its available thermonuclear fuels can no longer support the star. Through a complex process that we are only now beginning to understand fully, the entire star explodes and ejects nearly all of its mass into space in a powerful, high-velocity shock wave. These supernovas occur mostly in the arms of spiral galaxies about once every 50 years. The last one observed near us was

SN1987A in 1987, located in the Large Magellanic Cloud some 160,000 light-years away and probably shouldn't count for the Milky Way's quota. We might be surprised by a supernova in our lifetime.

190   Will Sirius ever become a supernova?

About 8.3 light-years from the Sun, the star Sirius dominates the winter sky. This star has a dense companion "white dwarf" (mass = 1.03 Suns) that orbits the massive primary star (mass = 2.14 Suns) at a distance of 13 AU every 50 years. Since it is known that Type I supernovas are produced in binary star systems, it makes sense to worry about Sirius someday exploding in our neighborhood. We do not really know exactly how some white dwarf systems become Type I supernovas, but the minimum requirement is that the normal companion star cannot be too far away from the white dwarf and that it has to be an evolved red giant star shedding mass. Some of this mass is collected by the white dwarf pushing it over the stable mass limit of 1.4 solar masses for a white dwarf. It then collapses into a neutron star or a black hole producing a supernova. Once Sirius A becomes an ordinary red giant in a few hundred million years from now, its outer layers will probably not extend more than a few AU. Sirius B will be well outside the atmosphere of Sirius A. I do not see how this would be favorable for a supernova, but it might be a nova from time to time. Good for a few flare-ups every few hundred years or so!!

191   Can astronomers predict when a star will become a nova or a supernova?

Novas are produced in binary star systems in which one of the two stars is a white dwarf or a neutron star, and the other is a normal star, usually a red giant that has a distended outer atmosphere. Some of this material can be captured by the dense companion. Depending on a variety of factors, the dense companion will flare up as a nova periodically, or simply blow itself to pieces as a Type I supernova. Astronomers know of many reoccurring novas; some work with near clockwork precision and their outbursts can be roughly predicted. The dynamics of the nova process are sufficiently complex that we as yet have no good way to predict when outbursts will occur from a cold start. Recurrent novas and dwarf novas are classes of stars that produce very large nova-like outbursts, usually involving instabilities in the accretion disks orbiting white dwarfs in binary systems. They can have outbursts every 500 days or less but are not periodic like clockwork so that each pulse can't be predicted with much accuracy beforehand. Supernovas are explosions that detonate the entire star. Some of these are nova-like situations such as the Type I event above. Many others are the so-called Type II events that involve the detonation of a single massive star as it ends its life. We have good theoretical reasons to believe that supernova precursor stars will be red supergiants if their heavy element content is like that of the Sun, or perhaps blue supergiants if they have somewhat more heavy elements, as was the case for Supernova 1987A in the "metal-rich" Large Magellanic Cloud. Just before it becomes a supernova, a star may have a composition that suggests a great deal of chemical enrichment of its surface by convection from its nuclear core. There may be a sudden increase in the velocity of its stellar wind or the thickness of its circumstellar dust envelope. As we learn more about the theory behind the explosion, we can better anticipate just what kinds of things to look for. Right now, we can identify which star type will explode, but not when this will happen to within a million-year accuracy or worse.

192   What is SS–433?

It is an unusual object located 16,000 light-years away in the constellation Aquila. In the 1980's, this object was all the rage in the astronomical community. It was a star with an unseen companion, which has an accretion disk orbiting it that was shooting out jets of gas at nearly the speed of light. Astronomers were very excited that, within our own galaxy, we had an object that resembled many of the most distant quasars and other radio sources, which also seemed to be doing this trick. Basically, it was believed to be a neutron star in orbit around a very luminous O-type star. The neutron star was dragging material out of its companion, creating an accretion disk around itself. Through magnetic forces, the disk plasma squirts out along the rotation axis of the disk at about one-quarter the speed of light. Further studies of this system led astronomers to eventually upgrade the neutron star to full black-hole status. Chandra X-ray observatory studies of this system announced in February 2003 were able to detect two lobs of 50-million-degree gas one light-year apart on either side of the black hole. The detection of this gas so far away from the system is very exciting because there has to be some way to keep it hot at such a great distance. Long-term monitoring has now shown that the black-hole accretion disk ejects bullets of hot plasma that travel at one-quarter light speed and probably rear-end the distant slower-moving material, reheating the plasma to millions of degrees in the "train wreck."

193   How hot can a normal star get?

The hottest known stars that are not white dwarf cinders are the Wolf-Rayet stars. These are massive stars (more than 20 times the Sun's mass) that have temperatures as high as 90,000 F. They are enormously active stars that are literally trying to blow themselves apart with powerful stellar winds. The central star in the planetary nebula NGC 2440 is currently on record as this record holder with a temperature estimated at 360,000 F. Astronomer Sally Heap at the NASA Goddard Space Flight Center used the Hubble Space Telescope's Planetary Camera in 1995 to zoom in on the core of the nebula and resolve this point-like white dwarf star at its very center. Located 4,000 light-years from Earth in the constellation Puppis, this is one of the hottest white dwarfs known. The white dwarf star H1504+65, discovered in 1986, also has a temperature similar to NGC 2440 and is the seventh brightest X-ray source in the sky.

194   Where did all the heavy elements come from?

According to big bang theory, the only elements that could emerge from the actual formation of the universe are hydrogen, helium, lithium, deuterium, and beryllium, so where did all the other 90+ elements come from? Astronomers have known for decades that as stars evolve, the nuclear reactions in their cores transmute lighter elements into heavier elements. For most stars like the Sun, this steady enrichment and production of nuclear "ash" leads to elements up to carbon. For stars somewhat more massive than the Sun, core temperatures are hot enough that even carbon can fuse with all the other elements lighter than itself to produce the elements nitrogen, oxygen, neon, sodium, and magnesium. In still more massive stars, even heavier elements will be able to fuse to produce energy, but once you get to the point that you are producing iron as an end product, the fusion process can no longer produce more energy than it uses in assembling the nuclei. At this point, the star will explode as a supernova. In fact, for all stars more than about six times the mass of the Sun, you end up with a supernova. In all cases, most or all of the star's mass will be expelled back into space, including all of the material, which has been processed by fusion into elements up through iron. But even during the brief minutes of the explosion, a tremendous processing of some of the material will occur that results in the production of all the other elements heavier than iron. Many of these are short-lived isotopes and barely make it back into the interstellar medium before they vanish altogether. But others, such as all the stable elements found on Earth, do survive and then contribute to the next generation of star formation. Have we actually seen this happen? In February 1987, a supernova exploded in the Large Magellanic Cloud, and it instantly became a test bed for the most sophisticated models we have about nucleosynthesis. It didn't take very long before astronomers were able to detect the elements cobalt, nickel, iron, and titanium, whose radioactive decay produces the fading light from the supernova. The amounts of these elements that were detected matched the computer models for the nucleosynthesis process. Although oxygen makes up only 1 percent of the interstellar medium, about two solar masses of oxygen were produced and ejected by SN 1987A into the interstellar medium—a net increase of 2,000 percent. So, astronomers are very confident that they understand how elements are produced in the cosmos since the big bang.

195   If you could travel to any star cluster, which one would be the prettiest?

The Jewel Box (NGC 4755 at 7,500 light-years) in the southern hemisphere is a star cluster about 20 light-years across with 198 stars. The stars from Earth are exceptionally pretty with colors varying from red and yellow to the more common blue-white with a number of luminous B-type giant stars and a red super giant star. It is also situated near the dark nebula called the Coal Sack. A planet orbiting one of the member stars would have a night sky blazing in brightness with hundreds of stars brighter than Venus or Jupiter at their most brilliant. Several stars may even exceed the brightness of the full moon if we were close enough. These would appear as a dazzling red and several blue stars. The Arches star cluster in the center of our galaxy is even more spectacular and deadly. It contains 150 of the most massive O-type stars crammed into a region only one light-year across, with powerful stellar winds that constantly blow through this volume. An unfortunate planet living in this stellar hotbed would find 150 blue-white stars scattered across the sky about 20 degrees apart, and each shining with the brilliance of the full moon. The stars would start to go supernova every few million years for a stellar light show unsurpassed anywhere in our galaxy.

196   Are there any civilizations that never paid much attention to stars?

It is very hard to study this question because the vast majority of the civilizations that have existed so far have not left enough records of their particular ways of thinking about nature to fairly reach a conclusion. All of the major civilizations have had a very strong interest in the stars, going so far as to recognize and name dozens of constellations, and develop stories for them. They differ only in the degree to which they thought quantitatively about the sky. The Babylonians recognized the planets and stars and calculated detailed rising tables for the bright planets. The Egyptians were not very accurate in reproducing even the most prominent star patterns on their tomb ceilings, but they recognized the importance of Sirius as a Nile flood-forecasting tool. The hardest civilizations to track are those for which oral record keeping was dominant. This includes most native African nations and many North American groups. Even the Vikings left behind few records that indicated a knowledge of more than the North Star. The Incas had an entire mythology based on what they called the Dark Constellations—those black blotches that run through the Milky Way. They didn't seem to pay much attention to the stars themselves, at least not the way that ancient Greeks, Babylonians, and Egyptians did.

197 Have we ever seen two stars eat each other?

This actually happens quite a lot in binary, two-star systems where the stars formed close together. The way that stars in a binary system evolve is complicated by the fact that although the orbits may stay the same for billions of years, the sizes of the stars change as they evolve. If stars are so close together that one engulfs the other, this can cause enormous friction between the two stars, causing them to drift together over time. Astronomers have seen many of these "common envelope" stars, but generally they do not last long. Eventually the turbulence caused by the dense cores sweeping through this envelope ejects the envelope into a nebula. One such system is PG 1114+187 with a period of 1.76 days. The short orbital period also leads some astronomers to the conclusion that the system passed through a common envelope phase at some time in the past. As the two stars orbit, they lose orbital angular momentum and energy to the envelope, and may eventually eject it as in the case of the planetary nebula M2-9 studied in detail with the Hubble Space Telescope. The "butterfly wing"–shaped nebula NGC 2346 is about 2,000 light-years away from Earth in the direction of the constellation Monoceros. At the center of the nebula lies a pair of stars that are so close that they orbit around each other every 16 days. This is so close that, even with Hubble, the pair of stars cannot be resolved into their two components. One component of this binary is the hot core of a star that has ejected most of its outer layers, producing the surrounding nebula. Astronomers believe that this star, when it evolved and expanded to become a red giant, actually swallowed its companion star in an act of stellar cannibalism. Ten years after Supernova 1987A in the Large Magellanic Cloud became the first naked-eye supernova in our skies since 1604, Oxford astronomer Philipp Podsiadlowski says that the mystery about the nature of the star that produced this extremely unusual supernova is probably solved. The star that exploded used to be a member of a close binary pair, but it merged with its partner in an act of stellar cannibalism in the relatively recent past—perhaps some 30,000 years ago. A significant fraction of the material from the core of the star seems to have been thoroughly mixed with its outer layers and the complex nebula surrounding the supernova remnant. Only the turbulent conditions in a common-envelope binary could have kicked up the gases to this extent. During the merger, which itself takes perhaps only a few weeks to complete, the companion star is completely destroyed and its material is mixed with the envelope and part of the core of the progenitor. It now seems that a double-star merger scenario is the only way in which all the various anomalies of Supernova SN 1987A can be understood with as few loose ends as possible.

198   How can an ancient white dwarf orbit a star only 100 million years old?

In binary star systems, the stars often are close enough that they can exchange mass. Because the masses of the stars are generally not exactly equal when born, the more massive star will reach the red giant stage first, and will shed mass into space through powerful stellar winds. This mass can be captured by a companion star, and over time it can "reset" the clock of the companion star, which will now behave as a massive star. The red giant then evolves into a white dwarf star, but its companion is now a star that is so massive it may run through its remaining life in only a few hundred million years. It is estimated there are some one million close-binary star systems in our galaxy in which there is transfer of mass from one star to the other. About 1,000 of these have been catalogued. Sirius is an example of just such a system where an A-type star with an age of about 500 million years is orbited by a white dwarf star, which is usually the remains of a star that has taken billions of years to become a red giant star before leaving a white dwarf remnant behind.

199   What are rare-earth stars?

For over 100 years, astronomers have classified stars according to the intensities of specific lines produced by various elements such as hydrogen, helium, and calcium. This led to a star classification sequence O, B, A, F, G, K, and M, which ranges from the hottest O stars (90,000 F) to the coolest M stars (6,000 F). A-type stars have temperatures near 20,000 F and have very simple spectra, but some are surprisingly complicated for stars that are not supposed to be cool enough to have convecting gases near their surfaces. The Am—magnetic A—stars are known to have magnetic fields 100 and even 10,000 times stronger that our Sun's field. These fields may not cover the star's entire surface but may be concentrated like huge sunspots across much of the star's surface. Another type of A-star are the Ap "peculiar A-type stars," which show unusual abundances of some very odd elements such as europium. One of the strangest of these is Przybylski's Star, also known as HD 101065, an 8th-magnitude star in the constellation Centaurus. When the light from this star is analyzed with a spectroscope, instead of finding the familiar spectral line fingerprints of hydrogen, calcium, iron, or other "light elements," astronomers detect a bizarre mixture of rare elements instead. Lines from elements with unfamiliar names such as holmium, samarium, dysprosium, and neodymium clutter the spectrum. These lines are so strong, in fact, that those from the more familiar elements cannot be seen at all.

200 From Alpha Centauri, where would the Sun appear to be in the sky?

From Alpha Centauri's current coordinates at Right Ascension = 14h 39.6m and Declination = −60d 50', the Sun would be located at Right Ascension = 2h 39.6m and Declination = +60d 50'. This places it in the constellation Cassiopeia as a bright first-magnitude yellow star (like Alpha Centauri) just north of the Double Cluster in Perseus and near the nebula IC 1805. Astronomers would quickly notice that it has large planets because its path along the sky would jiggle in a 12-year cycle by about five seconds of arc. Alpha Centauri doesn't do this, so we conclude that it has no large planets in orbit around it.

201 How many planets do you think there are with intelligent life?

This is a classic question in college introductory astronomy courses. Students are asked at some point to supply their own estimates based on what they have learned about astronomy. They usually have to estimate the number of stars that have planets; the fraction of planets in a temperature zone where liquid water could exist; the fraction of planets that ended up with bacterial life; the fraction where this life evolved into multi-cellular organisms; the fraction of planets where multi-cellular life became technologically sophisticated; and the lifetime of that civilization. Astronomers often get numbers between one civilization and one million. We know that in a galaxy such as ours, the answer is one at least, but until we start seriously listening for others using programs like the Search for Extra-Terrestrial Intelligence (SETI), we will never know if it's much greater than one. The SETI program uses radio telescopes to listen for intelligent radio signals from thousands of stars. Right now, after listening for several decades, we know that the odds for other civilizations living right now are probably no higher than one in a hundred. By the time we get to one in a billion, we will have arrived at a definite answer to this question without having invested one penny in developing interstellar travel. I think that programs like SETI are a bargain no matter how you do the math.

202   When were "nebulae" first discovered?

Actually, humans have seen nebulae for thousands of years. They just didn't bother to single them out by name. Only the Pleiades and Praesepe star clusters seem to have a history that goes back more than a few thousand years. After Galileo first started using the telescope in 1609, the sky opened up for scrutiny and several generations of astronomers began the tedious but exciting job of cataloging what they could find. Nebulae were discovered soon after serious studies of the sky were begun in the 18th century by Charles Messier (1730–1817), who was searching for comets. He would often stumble across faint smudges of light, which he identified in a catalog of some 100 objects you should ignore as "not comets." His first entry was M1, the Crab Nebula, in 1758. No great interest in nebulae was taken until 1783, when William Herschel (1738–1822) began assembling a catalog of over 2,500 nebulae. By 1864 his son Sir John Herschel (1792–1871) had assembled a *General Catalog* of over 5,000 nebulae, and J. L. Dreyer in 1888 produced the *Index Catalog* with over 13,000 entries.

203   On a clear night I saw Sirius changing colors from blue to white. Why does this happen?

This is just refraction of the various light rays from Sirius by the turbulent atmosphere. Technically, it's called spectral dispersion. Think of the image of Sirius as a series of overlapping images at each color of the visual spectrum from blue to red. When the air is stable (which it never is!), the images overlap and you see what looks like white light. Depending on its density, water content, and dust content, the atmosphere scatters the short wavelength light more severely than the longer wavelength light, so the overlapping color images of Sirius get smeared out and you see both the individual twinkling effects and color simmering as the atmosphere shifts the various colored images around.

204 Do magnetic fields exist throughout space?

Yes there are magnetic fields in space, but their strength depends on where you are. The spiral arms of the Milky Way seem to contain a very large organized magnetic field. The spiral arm field is not much stronger than 5 millionths of a Gauss (5 microGauss) and seems to be wrapped around the arms like a Slinky. Earth's magnetic field is about 1 Gauss, but 5 microGauss over something as large as the Milky Way represents a lot of energy. Interstellar dust clouds have been found with fields as high as a few milliGauss—about 1,000 times stronger than the Milky Way's average magnetic field. As these clouds collapse, the fields are amplified to stellar and planetary strengths. On the cosmological scale, there are no data to suggest that magnetic fields are present, except in the vicinity of objects such as quasars, radio galaxies, or other kinds of active galaxies. As these galaxies eject clouds of gas, the gas drags with it some of the magnetic field within the cores of these galaxies. Using radio telescopes astronomers see enormous magnetic clouds, which can be millions of light-years across and look like puffs of smoke being launched into the intergalactic void. Some astronomers have speculated that the space between galaxies may contain lots of this magnetic gas, but the amount is much too small to be of importance to the evolution of the universe. Because it is so dilute, it has a lot of trouble cooling down, and so it probably just sits there suspended between the galaxies, hot and invisible, like a puff of smoke trapped in a jar.

205 Who invented the stellar classification system OBAFGKM?

The famous spectral sequence of stellar types was developed by Harvard astronomer Edward Pickering (1846–1919) based on an original scheme using the alphabet sequence from A to Q. He employed stellar spectra obtained in the 1890's to distinguish stars that had similar spectra. Then, in the early 1900's Henry Draper (1837–1882) at Harvard University completed the famous *Henry Draper Catalog* containing the spectra of over 200,000 stars, which led to the refined Harvard Spectral Sequence of OBAFGKM, because many of Pickering's original classifications were found to be redundant or were swapped around to make a more uniformly changing sequence. In 1943, the OBAFGKM sequence was improved with decimal sub-types such as O4, G5, and M7 based on precise spectral line strength measurements of certain diagnostic indicator lines. A G5 star is a star with spectral features that are "halfway" between those of a G0 and a K0 star. In time, three more classes were added to allow for very cool stars R, N, and S. Then astronomers added the C class to identify carbon-rich stars. Also, these classes could be refined by adding subscripts such as e, p, or m, meaning that there were particular emission lines found in the spectrum, that the spectrum was in some way peculiar, or that the star had a strong magnetic field.

206 How many stars are there?

This is actually a hard question to answer because when you get right down to the science, the only thing that astronomers often care about when it comes to galaxies is their total mass, not their total stars. Table 5 gives you some idea of the many estimates for the mass and star content of our Milky Way. The total mass estimates tend to be near 700 billion solar masses. How do we convert a mass to a number of stars? From detailed surveys of the 33 stars nearer than 13 light-years from the Sun you find one A-type, one F-type, five G-types, five K-types, and 21 M-types, and two white dwarfs. The average masses of these stars are A=2, F=1.5, G=1.0, K=0.8, M=0.5 for a total mass of 23 solar masses. That means that for every 23 solar masses, you get 33 stars, or that the average mass of a "star" is 0.7 solar masses. So, if the Milky Way mass is proportioned the same way, 700 billion solar masses implies about one trillion stars for a rather large galaxy such as our own. To find out how many stars there are in the visible universe, we have to multiply this Milky Way population by the number of galaxies in the visible universe, and their average size relative to the Milky Way. Numerically, as in our particular cluster of 33 galaxies, most galaxies are "dwarfs" with typical masses about 0.01 to 0.001 of the Milky Way. The estimate from the Hubble Deep Field is about 80 billion galaxies, but in order to see them from where we are, they would have to be giants like the Milky Way. Added to this would be perhaps 20 times this number in dwarf galaxies. So, when we combine this information, there are about 80,000 billion billion stars in the visible universe—give or take a factor of 10 or so! The last time I calculated this number publicly, I got 500,000 billion billion stars, which is only a factor of six different! Astronomers really don't find this number very interesting, so they don't tend to invest a lot of time and trouble to get it right.

207　How many stars are born each year?

The Milky Way has been around for 12 billion years or so, during which time about 700 billion stars have formed. This means an average rate of 20 stars per year, but this is just an approximation. Most of this star-forming activity happened within the first five billion years of our galaxy's formation, because we can see lots of very old stars all around us. In some interstellar clouds such as the Great Nebula in Orion, about 500 stars have formed in the last 10 million years, and there are several thousand other star-forming clouds just like it that are active today. So the actual current rate may be closer to about a "few" stars each year in a galaxy like ours.

208　How close together are stars in star clusters?

The typical density of stars in the core of a globular cluster can be 100 or more per cubic light-year. The cluster Omega Centauri, for example, has a star density of about 180 per cubic light-year. This means that the typical distance between stars is about six solar system diameters. The night sky would have thousands of stars brighter than Venus, with possibly a few too bright to look at without severe retinal damage. Most of these stars will be red in color, but you may be lucky enough to be near a few of the mysterious "blue stragglers."

209　What are the colors of the stars in the Big Dipper?

Only the stars Dubhe (orange) and Alkaid (blue-white) would be expected to have any color based on their spectral classes of K0 III and B3 V. The rest of the seven stars are all A-type stars and would be white in color. Most stars will look white because they are so faint that they don't activate our eye's color-sensing cone cells.

210　Would the center of the Milky Way be hostile to Earth-like planets?

Yes, it probably would be over the lifetime of a civilization like ours. The interstellar medium in the galactic center contains 100 to 1,000 times the amount of gas in our neighborhood. Depending on your exact location, the cosmic radiation flows are also hundreds of times what they are near the Sun's distance. The solar system would be frequently invaded by dense interstellar dust clouds, which would reduce the amount of sunlight we would receive and probably bring about terrible ice ages. If the Sun passed through a dense cloud with about 10,000 atoms per cubic inch, the ionosphere would be eroded, and the ozone layer would disappear, rendering us vulnerable to harsh ultraviolet radiation. Also, on timescales of only a few thousand years, the stars in our skies would change their positions dramatically. Depending on the mix of stars in the core, you could find yourself next door to some rather nasty luminous O-type stars and supernova remnants. Every few million years or so, you would have a supernova explode within a few light-years of the solar system, which of itself would probably not be lethal with a thick enough atmosphere, but it sure would be spectacular!

211   Do stars ever lose mass after they are formed?

Stars can lose mass in a variety of ways. Our own Sun is steadily ejecting a "wind" of particles at a rate of one 100 trillionth of its total mass every year. At the present rate, this will not amount to more than perhaps the loss of an amount of mass equal to Earth over the lifetime of the Sun. As stars age, this ejection rate increases greatly. It can reach as much as the mass of a star like the Sun in a million years or less. This kind of star is evolving through its so-called red giant phase en route to producing a planetary nebula, or for more massive stars, even a supernova. Stars in binary systems can exchange mass between themselves over the course of years or decades, which can change the evolution of the member stars. If one star expands to become a red giant, some of its material can fall onto the companion star, causing it to gain mass and evolve more rapidly as a massive star. The most drastic way a star can lose mass is of course through a supernova detonation. In a matter of hours, as much as half of the mass of the star can be explosively ejected.

212   How did astronomers discover how stars were formed?

As you can imagine, there is nothing attached to a star that says, "I am 10 billion years old," or, "I was just formed 1,000 years ago." To get this kind of information, astronomers have to use logical deduction based on a collection of clues they have accumulated over the decades. To form a star, you require a more primitive condition where the matter is present in another, more distributed form. This describes interstellar gas clouds and nebulae almost exactly. By examining nebulae, astronomers find objects that are starlike but display many spectroscopic properties that suggest violent gas infall or outflow, as if material is flowing into a region of high density. These objects are commonly so deeply embedded inside the clouds that you cannot see them with optical telescopes but have to use infrared telescopes to detect them. One dramatic infrared picture of a dark nebula in Orion shows dozens of optically invisible, star-like objects glowing brightly in the infrared. After looking at many different clouds, astronomers see objects with different properties, and they can order them in a rough chronology. The invisible infrared sources eventually "turn on" and become full-fledged stars. In several nebulae we can see these infant stars, which astronomers call T Tauri objects. They are typically stars with about the same mass as the Sun, and partially embedded in nebular material, but showing tremendous surface activity, such as flares, powerful winds, and so on. After about 10 to 50 million years, these stars become far less active, and we think this is because the gas and dust in their fetal environment have been blasted away or evaporated. Again, the trick is to look at as many different nebulae as we can to find additional "stages" in this formation process. We know that for massive stars, the formation process is vastly different than for low-mass stars like our Sun. Astronomers also use the laws of gravitation and nuclear physics to interpret the data and to organize the observations into physically consistent evolutionary sequences.

213 What is a quark star?

This is an extreme type of neutron star that is far denser, and more compact in size—literally a handful of miles away from being a black hole. Erlend Ostgaard at the University of Trondheim in Norway described the current state of the art in figuring out what the pressure inside a neutron star looks like at densities above two billion tons per cubic inch. If densities about 20 billion tons per cubic inch can be achieved in the core, quark matter will appear and you can end up with a hybrid star. Neutron stars still contain more than 90 percent of their mass in the form of neutrons, but since neutrons are composed of three quarks, an even more extreme condition can happen when densities and pressures become high enough. Then, the neutrons in the core dissolve into a gas of quarks. From the outside, it looks like a neutron star. The quark-matter core allows the neutron star to be a stable object at about the same size of 12 to 16 miles across as ordinary neutron stars. The

PLATE 1. Colliding galaxies. In 3 billion years, the Milky Way and the Andromeda Galaxy will end a 12-billion-year dance, merging together to form an elliptical galaxy. Astronomers have seen many examples of these death dances play themselves out in the cosmos, such as this pair of spirals NGC 2207—IC 2163. (CREDIT: HUBBLE HERITAGE TEAM, NASA/AURA/STScI)

PLATE 2. Earth and moon from far away. This mosaic, taken by the NEAR spacecraft from a distance of 250,000 miles from each body, is not quite a true-life view of them in proper relative size, but it is nevertheless a unique perspective on this unusual double-planet system from space. (CREDIT: NEAR SPACECRAFT TEAM, JHUAPL, NASA)

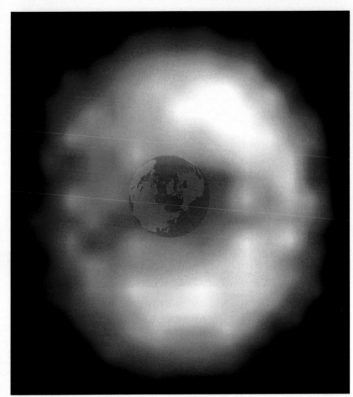

PLATE 3. An invisible river of charged particles orbit Earth like Saturn's rings. The IMAGE satellite can "see the invisible" by using special instruments. In this image, the ring current brightens as Earth's magnetosphere is pummeled by a solar storm. (CREDIT: IMAGE SCIENCE TEAM)

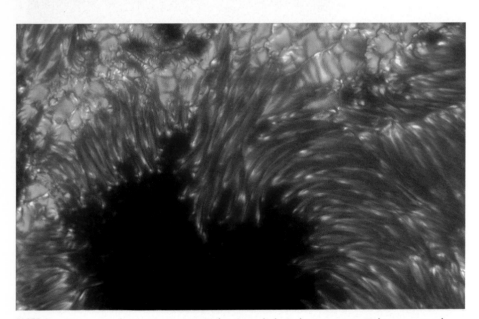

PLATE 4. Sunspot close-up. In 2002 the Swedish Solar Vacuum Telescope took this dazzling image of a sunspot. Details as small as a few hundred miles across can be seen. The magnetically heated gas flows out of the intense sunspot region (in black), whose dark recesses remain mysterious even with this technology. (CREDIT: ROYAL SWEDISH ACADEMY OF SCIENCES)

PLATE 5. Neutrino photo of Sun. Over 500 days of data were needed by Japan's Super Kamiokanda Neutrino Observatory to render this image of our own Sun. Although the resolution is still quite poor, as seen from Earth the neutrino-producing solar core would appear as a disk of "light" on the sky about 1/10th the diameter of the full moon. (CREDIT: R. SVOBODA AND K. GORDAN—LSU)

PLATE 6. Mars surface from Viking 1. I never get tired of looking at Mars this way. It seems so real, and so livable with its mysterious dark daytime sky, and its red-ochre landscapes fleeced with dust and stones. (CREDIT: NSSDC/VIKING 1 TEAM AND MARY A. DALE-BANNISTER, WASHINGTON UNIVERSITY IN ST. LOUIS)

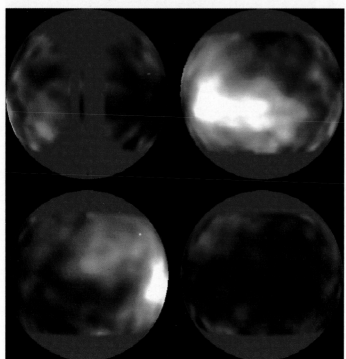

PLATE 7. Surface of Titan Four global projections of Hubble Space Telescope data separated in longitude by 90 degrees Upper left: hemisphere facing Saturn. Upper rig leading hemisphere. Lower left: hemisphere that never faces Saturn. Lower right: trailing hemisphere. (CREDIT: UNIVERSITY OF ARIZONA LUNAR AND PLANETARY LABORATORY; STScI)

PLATE 8. Map of Trans-Neptune bodies. The orbits of Jupiter, Saturn, Uranus, Neptune, and Pluto are shown, with Pluto indicated by a large white symbol. Over 600 objects are identified in this plot. Centaur objects (orange triangles), Plutinos (white circles), scattered-disk objects (magenta circles), main-belt objects (red circles), comets (light-blue squares). (CREDIT: BRIAN MARSDEN. IAU MINOR PLANETS CENTER.)

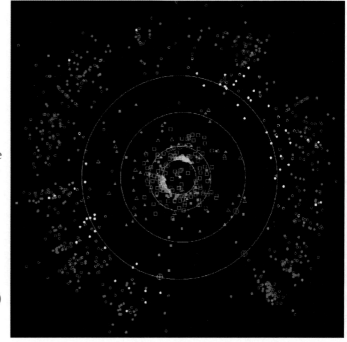

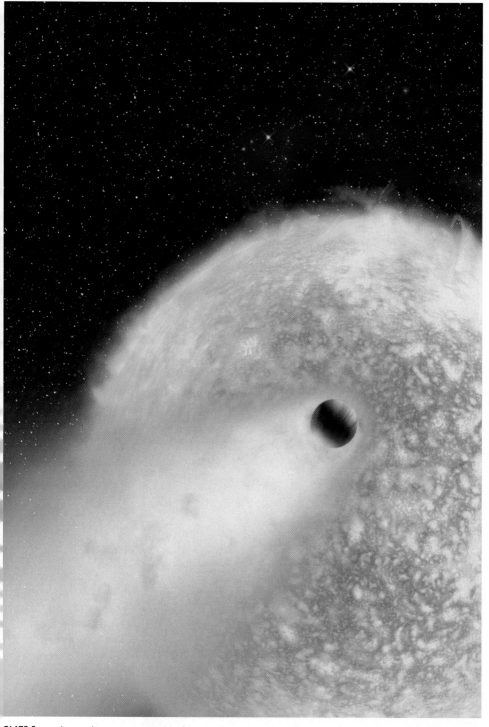

PLATE 9. Planet losing its atmosphere. An artist's rendition of the hydrogen gas cloud streaming from the exoplanet HD 209458b. In time, only the dense rocky core of this world will survive its scorching encounter with its parent star. (CREDIT: ESA, ALFRED VIDAL-MADJAR—INSTITUT D'ASTROPHYSIQUE DE PARIS, AND NASA)

PLATE 10. Planetary nebula MZ-3. An ancient star tears itself to shreds by ejecting its outer atmosphere. We still don't fully understand why many of these nebulae have such complex shapes. (CREDIT: NASA, ESA, AND THE HUBBLE HERITAGE TEAM - STScI/AURA)

PLATE 11. Milky Way's heavenly twin: NGC 1232. This beautiful spiral galaxy in the constellation Eridanus is located 100 million light years away. Many astronomers use it as an example of what our Milky Way might look like from outside. Our sun would be located about 2/3 of the way from the core region, in a spot perhaps not unlike what you would find in this image at about the 8 o'clock position. (CREDIT: EUROPEAN SOUTHERN OBSERVATORY)

PLATE 12. The first detailed, all-sky picture of the infant universe. This beautiful WMAP satellite image reveals the ancient temperature fluctuations (shown as color differences) that correspond to the seeds that grew to become the galaxies of our universe. (CREDIT: NASA/WMAP SCIENCE TEAM)

1 light-year

PLATE 13. The core of the Milky Way. The mysterious region surrounding the central black hole, SgrA*, contains many red giant stars and plenty of interstellar dust and gas. The arrows show the location of this black hole. (CREDIT: ESO/VLT AND REINHARD GENZEL MAX-PLANCK-INSTITUT FÜR EXTRATERRESTRISCHE PHYSIK)

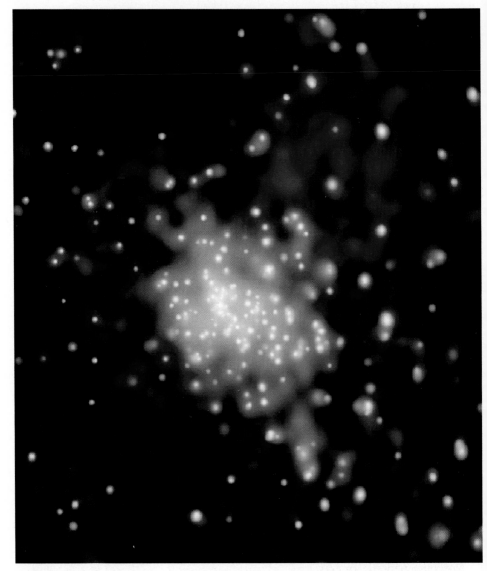

PLATE 14. The multi-wavelength universe. Star cluster RCW 38 is now viewable as a complete electromagnetic entity shining in the light from X-rays (red), radio (blue), and infrared (green). (CREDITS: X-RAY: NASA/CHANDRA X-RAY OBSERVATORY; INFRARED: ISAAC/VLT; RADIO: ATCA)

major difference, however, is that such a star would contain less mass: 1.4 solar masses compared to 2.5 solar masses. This is because their quark-rich cores provide higher pressure than an equal volume of neutrons and can support more mass. The mathematical structure of one such hybrid star shows a core of quarks, actually "strange quarks," out to a radius of four miles occupying 60 percent of the star's mass, with an outer layer three miles thick of neutron matter. Combining Chandra and Hubble Space Telescope data, astronomers found that RX J1856 has a diameter of about seven miles. This size is too small to reconcile with the standard models of neutron stars. It is one of the best candidates known for a quark star.

214   How do some stars make dust?

Among the more interesting red supergiant stars are those whose distended atmospheres (extending beyond the orbit of Earth) are so cool that carbon and silicon atoms can collide and get stuck together to form graphite and silicate dust grains. Stars that are oxygen-rich rather than carbon-rich prefer to form grains of silicon monoxide (SiO) instead. As more and more of these dust grains begin to form and condense out of the star's atmosphere, the light from the star begins to fade until in some cases the star becomes invisible. A bright infrared star called IRC+10216 seems to be one of these super giant stars, which has permanently enshrouded itself in dust so that we can now only see it by its infrared light. Some variable stars also make dust grains. As their outer layers expand and cool, dust grains form and further reduce the light reaching us. The change in the infrared light from these variable stars lags behind the optical light variation slightly because the dust grains do not form instantly as the stars' outer layers expand and cool.

215   What causes pulsating variable stars to pulsate?

Intrinsic pulsating stars are stars that, by some means, change their brightness in a periodic, or even irregular, way, over a span of hours to hundreds of days. The short-period variables are typically hotter stars in classes A through K, while the long-period variables are older, cooler stars of class M. Betelgeuse, in Orion, for example is a semi-regular, long-period variable whose brightness changes over a time of several hundred days. It is a red super giant star. Delta Cephi, the archetype for the Cepheid variable star, is a short-period variable with a period of 5.37 days. As a star evolves, its internal energy source increases, and the star's size also increases to accommodate the new internal pressure change. It continues to expand until gravity gains the upper hand. However, if the expansion is fast enough because the central energy source increased too quickly, then the inertia of the gas can cause the outer layers of the star to overshoot the stable point where gravity and internal pressure are in balance. The outer layers then slow down and fall back through the stable point until the internal pressure again spikes up, sending the outer layers back through the equilibrium point again. You would expect that this would damp out in time due to friction, but evidently there is an internal reservoir of energy that just compensates for the loss of kinetic energy due to friction. When helium is not ionized, radiation produced in the core of the star can pass freely through the interior of the star. During the star's expansion cycle, these conditions evidently prevail and the star brightens. During the collapse cycle, the interior is heating up, and eventually you reach the point where helium gets doubly ionized. Such a gas becomes opaque to some of the radiation attempting to pass through it. This causes radiation pressure to increase in the core as more of the light is forced into ionizing the helium atoms. A diminution of the star's light then happens, but eventually the process stops when this internal pressure becomes high enough. The collapse is replaced by a rebound, and as the star expands, the helium atoms gain back their second electron and the interior of the star is once again transparent. The cycle begins anew.

CHAPTER 6

The Universe— All That Is, Was, or Will Be

W e can dream about voyages to distant stars, and empires spanning our Milky Way, but not even the most talented science fiction writer can encompass the cosmos with a plausible human vision and saga. Each galaxy is a continent of star-cities separated by vast oceans of space and time. While our stars remain familiar and we privately may mourn the occasional supernova, what do we make of galaxies colliding and tearing themselves asunder? If there is but one intelligence in our Milky Way, are other galaxies graced by consciousness? Beyond the scale of the galaxy, our view takes in the billions of galaxies in our universe, seen through the awkward lens of the increasingly distant past. We live in a phantom universe, after all, where things that are seen are perceived as they once were. We learn as well that the destiny of the cosmos is not at all in the hands of the things we can see. It is shaped by the vast dark energies and matters whose natures we can only dimly fathom.

◐ ● ◑

216   What is the "anti-matter fountain" in the Milky Way?

In April 1997, scientists using NASA's Compton Gamma Ray Observatory announced that a fountain of gas was rising out of the center of the Milky Way. This wouldn't have been quite so unusual for a galaxy such as ours except that this gas consisted of anti-matter. As yet, no one really knows what this is all about. We do know that the core of our galaxy harbors a massive, three million solar mass black hole, and lots of very unusual gas and dust-cloud activity within the inner 100 light-years or so, but what process or event could have produced this plume of anti-matter? Anti-matter positrons are annihilating with normal matter electrons at an astonishing rate and producing the gamma-ray light detected by the satellite. One group of sci-entists favors the production of this anti-matter in the jet of the central black hole, while other scientists favor material produced by supernovas, flowing like a wind out of the Milky Way. "The antimatter cloud could have been formed by multiple star bursts occurring in the central region of the galaxy, jets of material from a black hole near the galactic center, the merger of two neutron stars," according to James Kurfess, head of the Gamma and Cosmic Ray Astrophysics Branch at the Naval Research Laboratory, "or it could have been produced by an entirely different source."

217   What is the Local Bubble?

The solar neighborhood out to 2000 light-years or so has several "bubbles" called Loop I, Loop II, and Loop III, each occupying its own roughly spher-ical volume in space, several hundred light-years across. Astronomers think these bubbles were formed by ancient supernova explosions that generated expanding shells of gas. We can see these other bubbles because the plasma trapped inside them emits X rays that our satellites can detect. Our own Lo-cal Bubble is about 200 to 300 light-years across, and is a low-density region populated by very tenuous clouds that you can't directly see with a telescope. Our solar system is entering one of these clouds called the Local Fluff. The properties inside the other bubbles may be similar, but the density of the gas may be higher if they are younger than the Local Bubble (about 100,000 years). It is all very complicated. Our Sun will eventually pass through many of these bubbles and pieces of "fluff" in the next few millions of years be-cause supernovas are rather common, and it is believed that the interstellar medium contains many of these old remnants. Entering one of these, we may find the flow of cosmic rays into our solar system increase for thousands of years before subsiding again.

218   Which galaxies are closest to the Milky Way?

The Sagittarius Dwarf Galaxy is closest, at a distance of only 78,200 light-years. Next we have the Large Magellanic Cloud at a distance of 159,000 light-years and the Small Magellanic Cloud at 189,000 light-years, followed by the Draco and Ursa Minor dwarf elliptical galaxies at distances of 247,700 light-years and the Sculptor System at 254,300 light-years. These are all clas-sified as irregular or dwarf galaxies and contain only a few tens of millions of stars each. The Large Magellanic Cloud is the largest of these and contains ten billion stars within a region of space about 40,000 light-years across.

219 What happens when galaxies collide?

Contrary to what you might think, when galaxies collide, it is very unlikely that even a single pair of stars would collide within the two galaxies. As the galaxies approach one another, the changing gravitational forces distort them, pulling spiral arms containing billions of stars out of the galaxies and sending the stars into completely different orbits. The collisions take over 100 million years, and the outcomes depend on the sizes of the two galaxies and the paths they take. Small galaxies colliding with big ones are usually eaten with little fuss. A head-on collision between galaxies that are roughly the same size, but where one is an elliptical and the other is a spiral type, can transform the spiral galaxy into a rare "ring galaxy." If the collision is off-center, the galaxies get completely shredded by their mutual gravitational fields. The end product may be a single new galaxy, or the two galaxies may continue on their ways, but with their new shapes very different from what they looked like before. Once again, Plate 1 shows what a collision between spiral galaxies can look like in real life.

220 What would the Milky Way look like from far away?

Because light takes time to get from place to place in the universe, from far away, you would see what the Milky Way looked like long ago. The image would show a galaxy younger than ours today. Also, because of the cosmological red shift, the Milky Way would appear to be receding from us at a speed determined by the Hubble Law. At a distance of 300 million light-years, the Milky Way would have a recession speed of 5,000 miles/sec. At three billion light-years, it would have a recession speed of 50,000 miles/sec. Its image would be that of a galaxy three billion years younger than our Milky Way's current age. Not much will have changed, but at in-between distances perhaps the core of our galaxy would have temporarily flared up as a Seyfert "active galaxy."

221 How do we really know what the Milky Way looks like?

Because we can't travel outside and look back, we have no choice but to search for clues and make educated guesses. We know it is a flat system of stars because on the night sky, we see most of the stars in a narrow band across the sky. From studies of other galaxies, those that are flat are also spiral-type galaxies with several arms. We know more or less where the major spiral arms are located, but details equivalent to photographic accuracy are largely unknown. Nearby, we know where some of the major star-forming regions, nebulae, star clusters, and large gas clouds are located. There are several competing models of what the Milky Way looks like from outside. The most recent idea is that we live in a spiral galaxy with a bar-like feature at its center, and with a small nuclear region. It would probably be classified as an SBb- or SBc-type galaxy in the classification system devised by astronomer Edwin Hubble (1889–1953) in the 1930's. If you want to see a galaxy that could be a near-twin of the Milky Way, have a look at Plate 11. This galaxy, called NGC 1232, is located 100 million light years away in the constellation Eridanus.

After nearly 50 years of intensive study, astronomers still can't figure out the exact shape of our galaxy in terms of how many arms this pinwheel-shaped galaxy has. We know that the Sun is on the edge of what is probably a spur that seems to merge with the Perseus Spiral Arm toward the constellation Cygnus. The Perseus Arm is the one just outward from the Sun's location in the Galaxy about 2,000 light-years from us. Beyond the Perseus Arm, there may be a more distant one, but the arms become less distinct in the outer galaxy zone. Interior to the Sun's orbit, which is about 25,000 light-years from the Galactic Center, and at a distance of about 6,000 light-years from the Sun, we encounter the so-called Sagittarius-Scutum Arm. Interior to the Sagittarius-Scutum Arm, there are indications of a Centaurus-Carina Arm at a distance of about 9,000 light-years from the center. Here we enter a very complex region where recent work seems to indicate that the Milky Way takes on a distinct "bar-like" geometry with a central core and two arms curling upon themselves like the Greek letter *theta*. Maps of the giant molecular clouds in the Milky Way seem to indicate an overall two- or possibly four-armed spiral bared-spiral shape, but there may also be many spurs jutting out from these main arms such as we find in other "grand design" spiral galaxies. Images of the entire sky taken in infrared light, such as the one by the 2MASS survey shown in Figure 15, give some idea of its shape as seen without any of the dust clouds to hide its stars. We may never know completely just what the Milky Way actually looks like. For now, we can trace a few of the major arm segments nearest the Sun with great fidelity, and in the distance see indications of how these segments join into a larger pattern.

222   Who discovered that the Milky Way rotates?

This is usually credited to the University of Stockholm's Bertil Lindblad (1895–1965) in 1926. In 1905, Jacobus Kapteyn (1851–1922) discovered that the proper motions of nearby stars were not random when the solar motion was subtracted from them. Instead they moved along a stream headed toward Lyra. Lindblad showed that this could be explained in terms of the

FIGURE 15 The Milky Way. Over a half-billion stars comprise this all-sky mosaic of the infrared Milky Way obtained by the 2MASS project. Stripped of its obscuring dust clouds, our galaxy clearly shows the flattened shape shared by all spiral galaxies, and gives us our only sense of this galaxy as it would appear from outside. (CREDIT: 2MASS/IPAC/NASA AND NSF)

rotation of the Milky Way galaxy itself. Astronomer Vesto Slipher of Lowell Observatory had also demonstrated a few years before Lindblad that other "external" galaxies also rotated; he did this by looking at their spectra, which were Doppler-shifted in a way that indicates rotation.

223 Do all galaxies have names?

Only the brightest and most interesting ones have names. These are usually just catalog numbers like NGC 5128 or 3C 273. This is as close as astronomers get to naming galaxies, and it only covers the fewer than 20,000 galaxies that are interesting enough for various research purposes. There are more than 50 billion galaxies in the visible universe that have not been "named," although many can be plainly seen on photographs. In terms of actual names rather than catalog designations, only about 120 galaxies have been singled out in this way. My favorites are Atom for Peace (NGC 7252), the Lost Galaxy (NGC 4535), and the Sunflower Galaxy (Messier 63).

224 Which way do spiral galaxies spin?

Spiral galaxies look like pinwheels in space with curved arms connected to a central hub. Once astronomers could measure the Doppler shift in the arms of nearly edge-on spirals, they could see that the arms trailed the motion of the rotation. This is the impression you get looking at a photo. But in 2002, Ron Buta at the University of Alabama used the Hubble Space Telescope to study the galaxy NGC 4622 located 111 million light-years away in the constellation Centaurus. Astonishingly, he found that its arms are leading. This galaxy is known to have been the site of some kind of galaxy train wreck or cannibalism "event" in its past, so perhaps that explains why it is such an oddball.

225   Are the spaces between galaxies filled with stars or gas?

Probably both. Intergalactic space, the space between galaxies, can be filled with a hot gas that emits X-rays, but this is only found within clusters of galaxies. Sensitive searches have been made for a variety of forms of hot and cold gas in the intergalactic space outside clusters of galaxies. Some indications are that, long ago, there once were vast clouds of hydrogen gas occupying these voids, but we do not know where these clouds went because they can no longer be detected except as absorption features in the light from very distant quasars. It is believed that many, if not all, of these clouds were ultimately eaten by galaxies. In addition to gas, some stars can get stripped from galaxies during collisions between galaxies. Astronomers are trying to determine just how many of these stars might be present, but it seems from the level of optical background light, that their numbers are rather low. It would be disturbing to live on a planet orbiting one of these. The night sky would be completely black, with only a few small smudges of light to stimulate our curiosity.

226   Does the Great Attractor still exist?

In 1986, Carnegie Observatory astronomer Alan Dressler and six other researchers found that the Milky Way and a host of neighboring galaxies are all meaning at hundreds of miles per second toward a spot in the constellation Centaurus, some 150 million light-years away. X-ray observations with the ROSAT satellite were able to identify a massive cluster known as Abell 3678 as the likely "core" of Dressler's Great Attractor. This cluster was previously known to astronomers as far back as the 1950's, but because it is located so close to the plane of the Milky Way, it was hidden by interstellar dust, rendering it very hard to see and making it difficult for astronomers to accurately count the number of galaxies within it. ROSAT has now uncovered a far larger cluster, and ESO telescope observations can now see a bewildering collection of galaxies in all stages of cannibalism and interaction, consistent with a massive supercluster of thousands of galaxies. Many of the galaxies in this huge cluster are slowly heading toward collisions with each other. The enormous galaxy cluster at the core of Abell 3678, called ACO 3627, shows the violent forces at work when galaxies get together. Twisted shapes and remnants of previous collisions show a continual process of galactic cannibalization. Does this contradict big bang cosmology, which says that all galaxies are drifting away from each other as the universe expands? No, because this theory expects that small regions less than a billion light-years across will have their own complex motions and concentrations of mass like the Great Attractor, but on still larger cosmological scales, all of these irregularities average out into a uniform patina of light and matter.

227   Do galaxies evolve?

Over time, everything changes. In the 1930's, Edwin Hubble studied hundreds of different galaxies and classified them into a "tuning fork" diagram with elliptical forms on the main shaft, and the two classes "spiral" and "barred spiral" on the arms of the fork. For a short while it was thought that by some means the very round elliptical galaxies slowly became flatter systems and changed into spiral galaxies. Once astronomers accurately measured how these systems rotated, they discovered that ellipticals contained very old stars and rotated very slowly, while spirals contained both old and new stars and rotated very rapidly. There was no known way for slow things to evolve by themselves into fast rotators any more than an ice skater on the ice can suddenly start spinning faster and faster from a standing start. By the end of the 20th century, astronomers began to recognize that galaxies are assembled from smaller clumps of matter, and even cannibalize their neighbors. As they do so, they accumulate more and more matter. Collisions between galaxies last hundreds of millions of years and can cause many different galaxy shapes to appear. Galaxies that used to be spirals, after collision and merger, can turn into elliptical galaxies, with powerful bursts of star-forming activity consuming much of the remaining gas and dust. So, galaxies evolve by frequent and complex collisions with their neighbors, but usually start out as smaller irregular-looking fragments of a few million stars. We have many of these irregular "dwarf" galaxies near the Milky Way, including the Magellanic Clouds and the Sagittarius Dwarf Galaxy.

228   When were galaxies first discovered?

Indistinct "nebulosities" were first discovered telescopically by Charles Messier in the 18th century, but he could not tell from their indistinct shapes exactly what they were, let alone their distances. By the mid-1800's, the major categories of the nebula had been identified, and some of the elliptical and spiral nebulae were being described as possibly outside systems of stars beyond the Milky Way. The recognition that galaxies, as we define them today, exist seems to be credited to William Herschel in 1783, when he noted, in his telescopic studies of nebulae, that "it may not be amiss to point out that some other very remarkable nebulae which cannot well be less, but are probably much larger than our own system; and being also extended, the inhabitants . . . must likewise perceive the same phenomena. For which reason they may also be called milky ways by way of distinction." By 1845, the Third Earl of Rosse (1800–1867), using a much larger, 72-inch telescope, was able to resolve stars in some of these nebulae. No one had a reliable way to measure the distances to these spiral nebulae to settle the issue of whether they really were outside the Milky Way, until Heber Curtis (1872–1942) discovered nova stars in several nebulae beginning in 1917. His distance estimate for some spiral nebulae exceeded three million light-years. The Andromeda Nebula was about 480,000 light-years, but it wasn't until Edwin Hubble's 1923 detection of Cepheid variable stars in the Andromeda Nebula star-swarms that the issue of the great distance of the island universes was finally resolved. The term "galaxy" did not enter the astronomical mainstream and replace "nebula" until the 1930's.

229 What is a quasar?

Quasars were first identified in the early 1960's, but aside from their great distance, no one had a telescope powerful enough to actually see what they were. They all looked like very bright points of light with absolutely no details to them. Theoreticians, meanwhile, were intrigued by the power they produced. By the mid-1970's, there were several ideas being proposed to explain quasars as massive black holes devouring entire stars and gas clouds and converting this mass into energy. These black hole models were later supplemented by "star burst" models where huge numbers of massive stars were going supernova—a much more common process than black holes that had not yet been proven. Several groups of astronomers in the early 1980's used the latest telescopes and photography to have a look at some of the closest quasars, such as 3C273. They reported detecting "fuzz" in their vicinity. Careful spectroscopic studies of this fuzz showed that it was at the same red shift as the quasar and that it looked like the light from spiral-type galaxies. Once the Hubble Space Telescope began to photograph the nearby quasars in the early 1990's, it found that nearly all of them were produced during galaxy collisions. One of these images, shown in Figure 16, gives a good idea of what the scientists typically saw. You can easily pick out several bright galaxy cores like the ones you find in the center of spiral galaxies, or you can see fragments of star clouds near the quasar. This led to the idea that a normal galaxy might have a supermassive black hole in its core, which is not eating gas and dust and stars. This black hole contains over 100

FIGURE 16 The mystery of quasars unveiled. This Hubble picture of the quasar PKS 2349 at 1.5 billion light-years proves that at least some, and possibly all, quasars are triggered by some dramatic event such as the merger between a quasar and a companion galaxy. The bright central object is the quasar itself, and looks like a brilliant star in the core of the galaxy—the home of a supermassive black hole feeding voraciously. (CREDIT: JOHN BAHCALL, NASA/PRINCETON)

million or even several billion solar masses of material and is about the diameter of the solar system. Then, along comes another galaxy. The collision tosses lots of matter into orbits that reach deep into the core of the host galaxy, where the black hole lives. The matter is absorbed by the black hole, which is like throwing gasoline on a fire. Our own Milky Way is constantly digesting small galaxies like the Magellanic Clouds, and it has a mid-sized black hole lurking in its central regions. So far it is not eating enough to turn on, but in three billion years it may turn into a full-fledged "active galaxy" spewing out powerful jets of matter and energy as the Andromeda Galaxy collides with us. If we had a billion-solar-mass black hole, we too would become a quasar!

230   What does the polarization of the cosmic background tell us about the Big Bang?

When you use sunglasses to shade your eyes, you are actually experimenting with an important property of light that tells you a lot about the surface that reflected it, or the medium it passed through to get to your eye. Scientists can measure polarization and use it as a diagnostic tool for studying distant gases and objects in the universe. Think of the cosmic fireball light as a powerful source of light, and the matter in the universe as a screen that you stand behind. By measuring the brightness changes in the light, you can figure out what the screen looks like. The polarization of the cosmic light tells astronomers that this light passes through a portion of this screen where the universe was clumpy and where regions of ionized gas had formed. Because young massive stars produce enormous regions of ionized gas, the polarization and its clumpiness are revealing the earliest times when stars were forming in the history of the universe. In December 2002, Erik Leitch and John Kovac of the University of Chicago and their colleagues used the Degree Angular Scale Interferometer (DASI) located in Antarctica to detect this polarization of the cosmic microwave background fireball. Hard on their heels, in February 2003 the WMAP mission announced its first complete all-sky map of this polarization. The WMAP scientists also calculated from its strength and clumpiness that the first stars had begun to form in the universe around 200 million years after the big bang. This is an important date because it established the end of the so-called Dark Era that began after 300,000 years when the universe had cooled and atoms became plentiful—but there were as yet no sources of visible light.

231 Will we ever be able to see what the big bang actually looked like?

The surprising answer to this is "Yes." When the universe first emerged from the big bang and underwent its inflationary episode, all of the sub-atomic variations in energy and space curvature imprinted at the big bang were magnified hugely. Today, by looking at the patterns coded in the cosmic background radiation, we can actually see a picture of these changes as though we were looking at the view screen of some enormous cosmic microscope. Astronomers have created predictions for what they should be seeing for a range of different "initial states," and it is hoped that in the next few years we might have the first image of what things looked like nearly 0.0000000000000000000000000000000000001 (10^{-35}) seconds after the big bang, when inflation started. Physicists using quantum theory and general relativity have also come up with a new way of thinking about the universe (see Question 263). Here's where the story gets real weird! The new "information-based" description of the universe says that if you want to describe what the beginning of our universe looked like in complete detail, it only requires about a gigabyte of information. The NASA WMAP mission will contain about this many spatial resolution elements in the picture it has now created of the cosmic background radiation shown in Plate 12. Coded in this image may very well be all the information we need to exactly specify the initial state of the cosmos.

232 What is the farthest object we know about in the universe?

Astronomers measure the distances to galaxies by referring to their cosmological red shift denoted by the letter z. This is a quantity that can be directly determined from the spectrum of the object. To relate this to a physical distance or a particular age for the object since the big bang, you have to know what geometric model of the universe you want to use. Table 6 gives a conversion from red shift to other physical quantities from the current standard model based on the latest WMAP observations for the cosmic expansion rate and its makeup of dark matter and energy. The record holder as of February 2003 was still the remote galaxy HCM 6A at a distance of 28.5 billion light-years (red shift z = 6.56), which was discovered in 2002 by Esther Hu and her team at the University of Hawaii's Keck II observatory. Thanks to some help from Mother Nature, they were assisted by the deep gravity well of a cluster of galaxies called Abell 370, which focused and amplified the light from this faint galaxy. It is one of the first known galaxies at a red shift greater than 6, and it was formed about 900 million years after the big bang, which makes it the youngest non-quasar object that we know about in the entire cosmos. Other contenders for this honor of Most Distant, as of March 2003, are also listed in Table 7. Just think. For this observation, astronomers used a telescope over 10 billion light-years long with a lens millions of light years in diameter, with the Hubble Space Telescope as the "eyepiece."

233 What is a gamma-ray burst?

For nearly 30 years, flashes of gamma-ray energy have arrived at Earth, invisible to ground-based instruments but quite apparent to satellite sensors. Their locations in space gave no hint of favoring the jumbled star clouds of the Milky Way. Instead, the nearly 3,000 bursts now recorded show a random pattern across the sky, as you can see in Figure 17. The bursts are also unique in their individuality as they silently flash in the night sky to their satellite observers. Through the coordinated efforts of NASA and ground-based investigators, in 1997 one of these fleeting explosions of gamma-ray energy was tracked back to its origin. The host for GRB970228 was a fading optical afterglow in an unremarkably faint galaxy about one billion light-years from the Sun. The afterglow from GRB971214 showed an object with the largest distance on record: 12 billion light-years ($z = 3.418$). If the light had flowed out in the manner of a normal supernova, the flash's brightness would have meant an equivalent power equal to nearly a billion galaxies like the Milky Way. It seemed that something far more powerful was needed to shine with such brilliance across billions of light-years of space. Calculations and computer models soon showed that just before the supernova detonation, core material with not enough energy to exit the doomed star would form a temporary disk located deep within the doomed star. As the star detonated, a powerful jet or beam of matter traveling at nearly the speed of light would punch its way out of the star along the axis of the disk—the path of least resistance. The jet would escape the star to cause the flash moments before the supernova begins its slow rise to maximum light. A cartoon sketch of the main parts of such an explosion is shown in Figure 18. A few of the famous record holders so far are:

- GRB971214, detected by the BeppoSAX satellite
- GRB 930131, the "Superbowl Sunday" burst detected by the EGRET instrument on the Compton Gamma-Ray Observatory
- GRB 021004, detected by the Chandra X-ray Observatory with the brightest afterglow seen so far, and
- GRB980425, detected by BATSE; the closest gamma-ray burst known so far, with a physical distance of about 120 million light-years

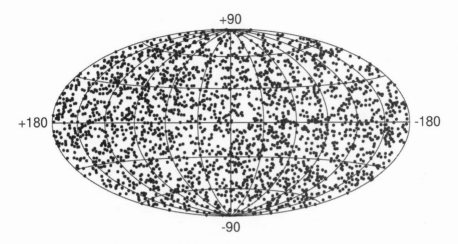

+90

+180

-180

-90

FIGURE 17 Gamma-ray bursts cover the sky in a completely smooth pattern. Once each day these cosmic flares flash on and off in different directions. Almost 3,000 have been charted to date. (CREDIT: CGRO/BATSE TEAM)

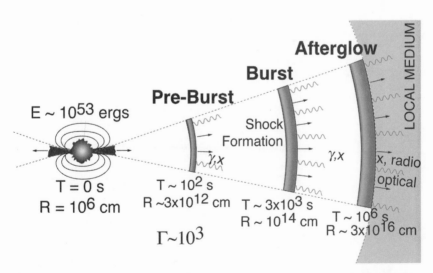

FIGURE 18 The anatomy of a gamma-ray burst. Thanks to a continued stream of data from ground-based and space-based telescopes, astronomers are narrowing in on a basic understanding of how some of these curious flashes get their start. (CREDIT: J. MYERS, GSFC)

234 What cosmic explosion lit up Earth in 1998?

For months, astronomers Chryssa Kouveliotou at NASA, Goddard, and her colleague Peter Woods at the University of Alabama in Huntsville had kept track of the fast-spinning neutron star SGR 1900+14, located 20,000 light-years away in the constellation Aquila. Then on August 27, 1998, at 5:22 A.M. (EDT), it suddenly let loose with a burst of energy. The gamma rays hit Earth's night side and ionized the upper atmosphere to a level seen only during daytime when it is illuminated by ultraviolet light from the Sun. It was so powerful that it blasted sensitive detectors to maximum or off-scale readings on at least seven scientific spacecraft in Earth orbit and around the solar system. The Rossi X-Ray Timing Explorer satellite's Proportional Counter Array was flooded with X rays even though it was not even pointing at this source. Data from the Ulysses spacecraft near the orbit of Jupiter showed radiation counts that rocketed over 1,000 times in intensity from normal levels in those distant regions of space. This object, called a Soft Gamma Repeater, was a known gamma-ray source that occasionally flared up to mimic the thousands of other gamma-ray burst sources seen by orbiting satellites. The major difference was that it was a repeating source located within the Milky Way galaxy, not at the far horizon of the visible universe billions of light-years away. The powerful burst seen on August 27 also proved that it was the first known example of a completely new class of astronomical object: a magnetar. These neutron stars have magnetic fields that are the most powerful in the known universe. If the moon were a magnetar, it would easily erase all magnetic recording media on Earth and rip the iron out of our red blood cells. At 1,000 trillion Gauss, the field is so strong that it actually alters the properties of space at the atomic scale.

235 What is MOND theory, and what makes it so interesting?

During the last decade, astronomers have had to confront the possibility that the universe is filled with dark matter and dark energy because this is the only conclusion one can draw from the data, and from Einstein's theory of general relativity, which is at the core of big bang theory. In 1983, an alternative theory for gravity was proposed by Mordechai Milgrom of the Weizmann Institute in Israel and called MOND (Modified Newtonian Dynamics). It resembles Einstein's and Newton's theories but also assumes that at very low levels—some 100 billionth the strength at Earth's surface—gravity behaves oddly, weakening more slowly than it does in standard theories. It sounds like a crazy idea, but the supporters of MOND point to dozens of predictions and calculations that seem to perfectly account for the missing mass in galaxies. They say that the theory is more than simply a lucky guess. Its successes have caused some astronomers to take MOND seriously enough to include it as an alternative to big bang cosmology. This period of open interest may now be near the end. A team of astronomers led by David Buote at the University of California at Irvine used Chandra to observe a hot halo of gas surrounding the galaxy NGC 720. Both the dark-matter and MOND theories can explain why the gravity of the galaxy is stronger than it appears. A critical difference between the two theories is their predictions for the shape of the halo: MOND predicts it to have the same shape as the galaxy, while dark-matter theory allows the halo to be shaped differently. What Buote's team found was that the shape of the halo was significantly different than the shape of the galaxy, contrary to the MOND predictions. For physicists, MOND theory is far more troubling. Up until astronomers began to find missing matter and dark matter in the cosmos, there was not the slightest indication that Einstein's theory of general relativity was missing something important. In its limit of weak gravity, it smoothly turned into Newton's physics of gravity, which solar-system astronomers and tide forecasters had used for decades and even centuries. MOND is not a description of gravity that can be turned into a theory that is compatible with general relativity because it requires that at the largest scales, gravity has to depend not just on your location but on your speed. No two observers will agree to the speed of an object according to relativity because there is no absolute speed reference frame. MOND theory is completely different than all other descriptions of physics that we know about and that work. This means that the stakes are very high for MOND, and so astronomers and physicists will continue to subject it to intense testing.

236 What is the youngest galaxy that astronomers have seen so far?

Astronomers Ester Hu at the University of Hawaii and Richard McMahon at the University of Cambridge used the new Keck Telescope to examine a galaxy at a current distance of 25 billion light-years (red shift of $z = 4.55$). It was located about two million light-years from a nearby quasar and is producing an enormous amount of Lyman-alpha light, which is a signature of the formation of very young, massive stars usually less than ten million years old. To see the light from an entire galaxy that shines by this light means there is a tremendous amount of massive star-forming activity taking place in this galaxy. The light from this galaxy shows us what it looked like when the universe was less than a billion years old. This is not necessarily the first galaxy to have formed in the universe, but it is probably a representative of a whole new population of infant galaxies we are now beginning to discover in which star formation was just starting. This makes them very interesting to astronomers, and I'm sure there will be more observations of such infant galaxies to come.

237 What is quantum cosmology?

For the time being, this is a highly speculative area of modern cosmology—the study of the origin and evolution of the universe. Quantum cosmology uses the tools being developed in the search for a quantum theory of gravity to address some very deep problems in cosmology. The most important and challenging of these questions is, What were the conditions like *at* the big bang? It also tries to explore new ways of thinking about how the universe got jump-started in the first place. Most of the discussions seem like the stuff of science fiction, or at least sound that way, but the practitioners feel that they are all on to something. In a few more decades, they may be able to tell us exactly what, and provide some experimental evidence for their assertions.

238 How did galaxies form from a smooth expanding big-bang state?

This is, perhaps, one of the most important problems in modern cosmology. In fact, all other problems can be considered, essentially, solved in modern astronomy, with this key problem the only remaining holdout! The big bang was a very smooth affair. We know this thanks to the results of many investigations and in particular the recent dramatic data from NASA's COBE and WMAP satellites. The universe is awash in what is called cosmic microwave background radiation—a relic from a time when matter and the radiation produced in the big bang were in intimate contact with each other. This means that the matter and radiation were physically interacting with one another, and the universe was opaque to its own radiation. This circumstance ended once the universe had expanded and cooled to a temperature of about 5,000 F, which occurred around 300,000 years after the Big Bang. Then, matter became neutral and non-ionized, and the fireball radiation saw the universe as a transparent void. COBE studied this radiation carefully for four years and discovered that to one part in 10,000 its intensity in the sky was the same in every direction. The material that came out of the big bang, at least at the large angular scales studied by COBE, was remarkably smooth, except for some very large-scale variations at a level of one part in 100,000 or so, which were over 100 million light-years across. COBE was not able to search for any irregularities that would be comparable in size to what we now see as galaxies. The WMAP mission, however, could study these irregularities, and in 2003 the WMAP scientists announced spectacular findings about the age and composition of the cosmos from even higher-resolution measurements of the cosmic microwave background.

Theoreticians studying the clumpiness of matter generated by gravity just *after* the universe became transparent to the fireball radiation. There are two scales of clumpiness in the density of matter that would have survived. The first contained clumps of matter equal to one trillion times the mass of the Sun; the second mass scale equaled about 100,000 times the mass of the Sun. The first is about the size of a very large galaxy or a cluster of galaxies; the second is about the mass of a globular star cluster. This material would have been a mixture of only hydrogen and helium, in the proportion appropriate to their cosmological abundances. There would have been no other elements. Astronomers are still debating when the Galaxy Formation Era began, but WMAP measurements suggest that the first stars were already forming about 280 million years after the big bang.

239   Did the birth of the universe violate the conservation of energy?

Yes it did. Very badly in fact. The reason this seems to fly in the face of common sense is that we have been told, mostly correctly, that energy cannot be created or destroyed. Once you measure the total energy of a system, that total stays the same as the system evolves in time. The problem we face with the universe is that this kind of definition doesn't work as well, when we have no foolproof way to actually calculate or measure this total energy. In Einstein's theory of special relativity, total energy is a well-defined quantity because spacetime is flat, and distant observers looking at your "system" will agree that at their location in time and space such a quantity could be defined uniquely. The problem is that near the big bang, spacetime becomes so convoluted and highly curved that there is simply no mathematical quantity you can construct in a flat spacetime that would be definable as the total energy of the system in a highly curved spacetime, or vice versa. This means that if you calculated the total energy of the universe today, at a time and scale for spacetime far removed from the big-bang conditions, you could construct a quantity called the total energy of the visible universe. If you tried to follow this quantity back in time, it would still be constant all the way back to the first second after the big bang when space was still mostly flat in geometry. However, if you tried to determine its value close to the Era of Inflation, or during the so-called Planck Era, spacetime is no longer deformable into a flat space upon which you can calculate an equivalent total energy. You cannot deform a cube into a sphere because it has six corners, and in a similar way, you cannot define a value for the energy of a system that doesn't include the shape of space occupied by the system, and its latent energy. The fact that energy was not conserved at the big bang is probably the controlling logical reason why the big bang got jump-started in the first place!

240 Have there been many major shifts in cosmological thinking?

In the 1920's, the biggest shift came in recognizing that the universe is expanding. In the 1940's, the next paradigm shift was that you could combine nuclear physics with big-bang theory to calculate the origin of the elements and predict the existence of a leftover fireball radiation. In the 1980's, the concept of "inflation" was added and resolved several long-standing problems with big bang theory having to do with the smoothness of the cosmic background radiation. Also in the 1980's, the growing conviction that we have not counted all the matter in the universe solidified into the notion that much, if not most, of the universe is in some form called missing mass, although it may not be matter-type particles at all, such as protons and neutrons. There have also been many smaller shifts having to do, for example, with the details of how galaxies and structure evolved in the universe. Today, the biggest shift is in the recognition that dark matter and dark energy really do dominate the evolution of the cosmos, and visible matter (stars and galaxies) is rather inconsequential in shaping the future destiny of the universe. Waiting in the wings once experimental evidence becomes compelling is the idea of the multiverse and other dimensions to space.

241 Where is the coldest spot in the cosmos?

It is usually the case that you cannot have something that is colder than its surroundings, and in a universe where the temperature of space cannot be colder than the cosmic background heat bath of 2.7 Kelvins, you would not expect to find anything colder. What two astronomers discovered in 1997 was an exception to this rule. After all, even an ice cube can survive in an oven for several seconds. A detailed study of the Boomerang Nebula, a nebula from a dying star located 5,000 light-years from the Sun, revealed that it was acting like a gigantic refrigerator to cool some of its out-rushing gas to a temperature of only one degree above absolute zero. Raghvendra Sahai of NASA's Jet Propulsion Laboratory and Lars-Ake Nyman of the European Southern Observatory and Onsala Space Observatory discovered that the winds were 100 times faster than for most other planetary nebulae and that the expanding gas was cooling parts of the nebula below the temperature of the rest of the cosmos. Once this wind abates, however, the laws of thermodynamics will demand to be repaid and the nebula will once again warm up.

242 How could a quantum vacuum fluctuation produce a universe with well-ordered laws of physics, rather than complete randomness?

We don't know. It is not known just how or where the laws of physics are ultimately proscribed for our universe. One possibility is that there was/is an infinite number of alternate ways that our universe could have emerged from this quantum process. In some sense, all of these alternate versions were/are realized, and our universe is one of a large number that were imparted with the right laws to continue the big bang and the evolution of our particular universe. Of course we do not know why there even are natural laws in the first place, but we do know that increasing temperature and energy can seriously alter the way these natural laws behave, and also the properties of the particles that they interact with. The natural laws at the beginning of the big bang may not have looked anything like the ones we now see, if you believe the hints that seem to be coming from high-energy physics experiments. Some physicists in the last ten years have speculated that at sufficiently high energies, all is complete chaos, and that natural laws emerge from this state in the same way that ice emerges from water that has cooled low enough. (See Question 259.)

243 Is it really true that the laws of physics were randomly selected at the big bang?

We honestly do not know. Obviously for this to be a scientific statement, we will have to be able to make some kind of observation to test this proposition. No physicist or astronomer has any clue how to test such an idea; so, many physicists and most astronomers view these kinds of statements as currently beyond science. This is a polite way of saying that the idea is not scientific because it is not falsifiable, no matter what its formal basis might be either in quantum mechanics, quantum field theory, or quantum gravity theory. Still, some theoreticians have discovered that very complex behavior emerges from simple elements operating under a set of simple rules. Complex phenomena seem to emerge from very simple rules. There have been mathematical toy models of universes created in which no identifiable physical laws are present, not even the principle of relativity, yet upon cooling to low temperatures, something like the laws of relativity emerge nonetheless. If the big bang was like that, and it certainly seems to have had a high enough temperature, then one worries just how inevitable our kinds of physical laws really are in the grand scheme of things.

244 Does the universe really have extra dimensions?

Because gravity and its field (called spacetime) are so intimately linked, asking whether the universe has more dimensions is the same as asking whether gravity works in a way that is consistent with its being exactly a four-dimensional field with three of these as the familiar dimensions of space. General relativity predicts that gravity with more than four dimensions will appear to be a very different force that will not obey the famous "inverse-square" law. Physicists have looked for traces of extra dimensions by doing lab experiments to test whether gravity is consistently an inverse-square force. In 1998 physicists Nima Arkani-Hamed at the Stanford Linear Accelerator Center in California, Savas Dimopoulos of Stanford University, and Gia Dvali of the International Centre for Theoretical Physics in Trieste, Italy, wondered how large the extra dimension could be, and how gravity would act if they were millimeters or millions of miles in size. A popular theory said that our universe was a 3-d "brane world" (see Glossary) adrift in an 11-dimensional space that gravity naturally operated in. They discovered that the more brane dimensions our universe had, the smaller would be the scale where departure from inverse-square law would be detected. In 2001, physicists Eric Adelberger and Blayne Heckel at the University of Washington announced their findings based on a careful study of the gravity-force law at millimeter-scales. If these extra dimensions existed, they must be smaller than 0.2 millimeters. It was a revolutionary experiment because up until then, no one had the technology to measure gravitational forces at this small scale. In another test of extra dimensions, physicists used the Fermilab Tevatron to search for "strong gravity" between particles involved in collisions of the D0 particle. If extra dimensions existed at scales of 10^{-19} inches, the result of these collisions would be slightly different. No such effects have been seen. So, this seems to mean that there are no extra dimensions larger than this minuscule size. But the hunt is not completely over, as physicists probe ever smaller distances at higher energies in future experiments.

245 What is the anthropic principle in cosmology?

This is an idea that many physicists and astronomers seem to be drawn toward to explain why the physical world has the particular properties that it does—properties that cannot be derived from the theories we currently have in our possession. In the future, some better theory may be crafted that can be tested, explaining what current theories do not, but for right now we don't know how to build them. For example, if the force of gravity didn't follow an inverse-square law, which must be the case for spacetime with exactly four dimensions, there would be no stable planetary orbits on which life could emerge. So the fact that spacetime is exactly four-dimensional is considered one of these features that we currently can't explain using general relativity. It seems to be a number you "dial in" to the theory and isn't defined by it. Also, the particular imbalance between the way matter and anti-matter decays has been measured in our labs as one part in 357. If it were only slightly different, there would be a lot less matter in the universe to make stars and planets. This also doesn't seem to be a number we can "derive" from any current theory that we have confidence in. Lastly, dark matter and normal matter form a ratio of 6 to 1. If that ratio were smaller, there would be less normal matter, too. Also, if the energy difference between fusing carbon and helium into oxygen (7.1616 million Volts) were only 0.6 percent lower, all of the carbon in a supernova would fuse to oxygen and there would be no free carbon in the universe to create the organic molecules needed for life. The anthropic principle says, basically, that some of the things we measure are the way they are because we are here to measure them with the values they have. It also says that life is intimately tied to the way the universe was set up right from the start. We are not accidents of nature at all.

246 What is time?

It is such a basic ingredient to the physical world that it is hard to identify its elementary properties. We all have a good mental picture of space, and the various things that occupy it from the human scale to the cosmic scale. To appreciate time you have to think in terms of timelines. Table 8 is the timeline for the cosmos and blends cosmic events with terrestrial and human events of the past, present, and future. It is this "fourth dimension" to organizing the universe that you have to think carefully about to understand why it is so important. We know that time is a different kind of "dimension" than the three we call space. Unlike space, we cannot navigate through time with any control, except that we can alter the way our brains perceive the passage of time. We can also manipulate the passing of time by moving, in the way that Einstein's special relativity tells us. We can measure time by using devices called clocks, but time is largely a synonym for change and has no other independent meaning. Nature does not allow us to visit the past and alter it, although we are free to visit the future as soon as we can get there. Time is an intimate ingredient to the gravitational field of the cosmos in the same way that space is. It is not something that exists apart from the cosmos in its own separate arena. Beyond these very crude facts, we know nothing about what time is of itself anywhere than we truly understand what space is. Some physicists such as Stephen Hawking have proposed that at some instant in the past "before" the big bang, time was actually a fourth dimension to space, but that by some fortunate but mysterious quantum tunneling process it became separated from the other three dimensions and turned into something different and distinct.

247 What did people in 1700 think the universe was like?

By this time, they knew the Sun was the center of the solar system. They also knew the size of the solar system to fair accuracy. Isaac Newton (1642–1727) had just published his work on gravitation in 1687, and Halley correctly predicted the return of Halley's Comet. By 1781 Uranus was discovered, Charles Messier made the first catalog of "nebulae," which were mistaken for comets, and William Herschel was busy mapping the stars and showed that the Milky Way is a flat system of stars with the Sun near the middle. The motion of the solar system through space was detected. Asteroids had not been discovered yet, and the universe was considered to be filled with stars, and infinite in physical size, thanks to estimates by Newton about the stability of matter under gravitational attraction. Nebulae were definitely the "mystery" objects of the time, and only Herschel is credited with suggesting that they might be distant Milky Ways. It was assumed that the Sun shone because it was made of some special material that just did that naturally. Some astronomers claimed to be able to see its darkened surface beneath the sunspots.

248   What did people in 1850 know about the universe?

We can get a good idea by having a look at some of the textbooks that were popular between 1840 and 1870. The massive 72-inch telescope built by the Third Earl of Ross was commissioned in 1845, and his spectacular drawings of distant nebulosities were slowly entering the mainstream of astronomical research into deep space. But it would be another decade before reports of his observations entered the general literature. The crucial instrument called the spectroscope was also not brought to the telescope until after 1850. The discovery that the Sun, planets, and stars contained familiar elements took a while to settle in. In some accounts, it was claimed that every fixed star was surrounded by a retinue of planetary worlds. The authors realized that the common-sense idea that brighter stars were therefore the closest was not entirely true. Stellar distances were routinely expressed in terms of millions of millions of miles. They were also aware that the solar system was moving at a speed of 28,000 miles per hour toward the constellation Hercules based on proper motion studies by Herschel. The origin of "shooting stars" was still entirely mysterious, though they were carefully being counted and studied. So far as the Milky Way is concerned, in earlier books, there was little mention. By 1870, Sir John Herchel's star-gauging work had provided grist for the mill of determining the shape of the flattened Milky Way system. Dark regions, called lacunae, were interpreted as directions where there were no stars. Star gauging resulted in a rather flat and very lumpy-looking Milky Way in 3-d. Telescopic data suggested that there were as many as 100 million stars to the depths of space that could be reached by these instruments. We see by this time the first speculations that the space of the universe might be infinite, but that the Milky Way may have only a limited extent in this vaster arena. A hierarchical structure occurs to some authors who proposed that nebulae that are resolvable with telescopes are within the Milky Way but those that are not are external galaxies. By 1873, books were already identifying aerolites, shooting stars, and meteors with matter entering the atmosphere from space. There were many accounts of aerolites hitting the ground and being recovered as stones. By some (incorrect) estimates using telescopes, more than 400 million of these small stones were encountered by Earth every night. The Great Nebula in Andromeda was an elliptical nebula resolved into thousands of stars. A distance of over 800,000 light-years had been estimated for it, though this unit was not specifically used. The Whirlpool Nebula in Canes Venatici was also seen as perhaps a system of remote faint stars like Andromeda. So, by the time of the American Civil War, other segments of society not on the battlefield or tending the farm had the luxury of contemplating the riches of a rapidly growing universe of stars and "island universes" made of familiar elements, and stars scattered through space like diamonds on black velvet.

249 Is the universe slowly rotating?

This is actually a very hard property to measure for the universe as a whole, at least the portion that we can see right now. Galaxies can only be detected to move along our line of sight. We cannot detect any possible movement across our line of sight, so a direct measure of rotation is impossible. Back in 1946, physicist George Gamov (1904–1968) proposed that the universe might be rotating because so many of the galaxies in the cosmos seem to be, but big-bang cosmology and observations put very firm limits to just how much rotation could be going on without making the universe look very different than the way we see it. If the universe were rotating, the abundances of the elements hydrogen, helium, and deuterium would have been changed during the first few minutes after the Big Bang. There would also be a change in the cosmic background radiation that would have been easily measurable by NASA's COBE satellite in the so-called quadrupole distribution of the background light. Even as early as 1973, physicist Stephen Hawking calculated that the cosmic background light would limit the rotation of the cosmos to about a trillionth of a rotation since the big bang.

250 How can you tell the universe is expanding from measurements made inside it?

This expansion of space can be measured for basically the same reason that you can tell you live on a round planet by doing some clever surveying within its surface. The main reason we can detect the expansion is that it is only a property of spacetime at the largest scales in the universe measured in tens of millions of light-years or greater. Regions of spacetime that are galaxy-sized or smaller are not, apparently, affected by cosmic expansion. This means we can define reliable yardsticks in our very small corner of the universe that can be applied to cosmological scales to detect the expansion effect via the cosmological red shifts we see.

251 Do atoms expand as the universe expands?

Careful measurements of the spectral lines from elements in remote galaxies show that over the last five billion years, the sizes of atoms have not changed from what we currently measure them. The energy transactions in atoms located billions of light-years away can be seen in their distinct spectral lines, and these show the same patterns and relative spacings as local atoms, so significant stretching at the atomic scale has not occurred. The cosmological expansion only applies to scales of several million light-years or greater. All smaller systems of matter are controlled by local, non-cosmological gravitational fields like those of the Earth, Sun, and stars.

252 Does the blue shift of the Andromeda Galaxy contradict big-bang cosmology?

No. The motion of a galaxy relative to us consists of the part produced by local gravitational influences, and a part determined by cosmological expansion. The typical speed of galaxies within their clusters, such as the Milky Way and the Andromeda Galaxy, is about 300,000 miles per hour. The cosmological expansion effect increases at the rate of about 454 miles per second per megaparsecs. This means that at a distance of three megaparsecs (9.8 million light-years), the galaxies' random motion within their clusters (say 120 miles per second) is about as large as its cosmological red shift. When you add these velocities together, you could get any speed from 0 to 250 miles per second. For nearer galaxies such as Andromeda, its random velocity is larger than its cosmological red shift so you can get a blue shift! This is why astronomers have to look at very distant galaxies so that they can easily see the cosmological red shift above the random speeds of the galaxies in their respective clusters. With enough galaxies in a cluster, these random speeds average to near-zero.

253 What is Olbers' Paradox and how does modern cosmology resolve it?

Heinrich Olbers (1758–1840) is credited with asking in 1826 why the night sky is dark. But it is very hard to believe that he was the first human to ask this astonishingly simple question. If we live in an infinite universe, which has been around for eternity, then every line of sight in the sky should end on the surface of some distant star. The night sky should be as bright as the surface of the Sun. This "paradox" is resolved in modern cosmology because the amount of space we can observe is not infinite, and stars and galaxies have not been around long enough for all their light to reach Earth. Also, because of the expansion of the universe, the light from the most distant galaxies gets shifted into other parts of the spectrum out-side the optical range, in the infrared.

254   Are we alone in the universe?

I think the chances are very good that life exists elsewhere. In fact, the prospects are far better today than they were ten years ago when astronomers couldn't even prove that planets were common in the universe. The planet Mars very likely had lots of running water on its surface when it was very young, but lost it all once the atmosphere leaked away. But there could have been a period lasting over one billion years when bacteria like that on Earth could have formed. That's why we are now excited about looking for fossils of ancient Martian bacteria. If we find them, that will mean that the condi-tions for evolving life, even just bacteria to start with, are fairly common and we could expect lots of bacterial life in the universe given that a planet orbits its star at the right range of distances. We now know that our solar system was formed in a way that is common to many stars. We have observed disks of matter orbiting young stars. We have detected planets orbiting nearly 100 stars. So, the chances are better than ever that there is life elsewhere in the universe. We still don't know how often evolution ends up with intelligent beings, though. This may be very rare considering the fickle conditions that had to occur on Earth for mammals and for intelligence to become an im-portant evolutionary skill. Dumb animals lasted billions of years on this planet, and survived five major Great Extinctions, before something like hu-mans evolved, and that was largely thanks to the dinosaurs' being forced into extinction by an asteroid impact 65 million years ago. A universe filled with bacteria and dinosaurs could be the rule, not the exception. So, we aren't alone so far as life is concerned. It's just that we probably don't have a lot of neighbors to talk to. As the universe gets older, though, this will certainly change. I am willing to wager that when the universe gets to be 100 billion years old, all of the possible accidents that trigger intelligence will have accu-mulated, and the cosmic radio babble will become intense. I would love to live long enough to test this hypothesis!

255   How long did it take an object now ten billion light-years away to get there since the big bang?

First of all according to big bang theory, which is based on Einstein's theory of general relativity, the distance between any two bodies in space may increase, not just because the bodies themselves are moving through the space but because space itself is stretching. This is like a runner on a wooden racetrack moving from the starting line to the finish line while at the same time the wood on a hot sunny day expands a bit during the race. Usually we don't include this effect because it is insignificant, and so it never enters into the stopwatch calculations we make. Cosmologists cannot calculate the distances to galaxies independently of the cosmological model that defines just how space is shaped and how it changes with time. The answer will actually be different depending on whether we live in an "open" universe that will expand forever, or a "closed" one destined to re-collapse. But all these big-bang models agree that a very distant object didn't get there by traveling through space from here to where it now is, like some fragment from a fireworks display. Instead, it is an independent clump of matter that was originally very far away from where we are, and the expansion of space has stretched the distance between us and that galaxy to where it is now. For example, in the case of the quasar PHL 957, astronomers measure its red shift and find that it has a value of 2.720. Thanks to NASA's WMAP satellite, we know that the universe is open, with an expansion rate of 44 miles per second per megaparsec, an omega parameter of 1.0, and a cosmological constant of 0.73, and that the current age of the universe is 13.7 billion years. With this cosmological model, we can now compute that we are seeing the quasar as it was 11.1 billion years ago when the universe was only (13.7–11.1) = 2.6 billion years old, and that its distance is now 19.8 billion light-years. Most of this distance has come from the space-stretching effect, which has nothing to do with the speed of an object through space. The matter that makes up that quasar was never mingled with the matter out of which our Sun and Milky Way arose, so it was never in any real sense close to us to start with.

256 If the Milky Way were the size of a penny, how big would the rest of the universe be?

A penny is about one inch across. If this equals 100,000 light-years as the diameter of the Milky Way, the Andromeda Galaxy is about 23 inches away, the Virgo Cluster of Galaxies is about 60 million light-years or 50 feet away, and the nearest quasar, 3C273, is about 1.5 billion light-years or a bit over 2 miles away. The visible universe has a horizon at about 13.7 billion light-years, and the most distant galaxies seen by the Hubble Space Telescope are nearly at this limit at about 12 billion light-years. This is about 20 miles away from the Milky Way penny, and is the farthest we can ever see. If the universe is infinite, then of course there is much more space out there in our model. Eventually we come to the planetary scale that represents the size of the cell that inflated to become our corner of the universe. Beyond the boundaries of this cell in our planetary scale model, we don't know what we would find. Perhaps a different kind of space, or perhaps nothingness.

257 Where does space end?

Big-bang cosmology and relativity say the universe is infinite. Locally, this means there is no edge to space. Astronomers only see and receive information from a small part of the universe we are actually living inside because light has only had 13.7 billion years since the big bang to do much traveling across the vast space of the present-day universe. We call this the "visible universe" just to keep in mind that the universe is actually much larger than the part we will ever be able to see. Our visible universe, however, expands at the speed of light, one light-year per year, so that billions of years from today, we will be able to see a much larger part of the entire universe that came out of the big bang. There is, however, no edge to space like the edge to a piece of paper. At least no theory we know today predicts such a situation! (See Question 36.)

258 How could the universe have been infinite even at the big bang?

This condition is predicted by general relativity, so the answer we have to find also has to come from within this theory. *At every instant,* the shape of the universe is infinite. The universe is infinite today. It was infinite when it was one minute old, and it was infinite *at the big bang.* A closed, finite universe cannot become an infinite one as it evolves, and a spacetime whose three-dimensional space-like portion is infinite today must have also been infinite when the big bang happened. This is demanded by general relativity. The key thing to remember is that if the universe is infinite now, as observations strongly suggest, then its three-dimensional volume *has always been infinite,* even a trillion trillion trillionth of a second after the big bang, when the local curvature at each point in this space was infinite. The caveat is that few physicists believe, now, that a purely classical description of this event is meaningful. That's why there is a massive search underway for a quantum theory of gravity, which will eliminate all singularities by smearing them out in spacetime and replacing them with a very large, but finite, curvature. The end result may be that "infinite" is just an idealization, and that what we really mean is something very, very, very big.

259 Is the order in the universe just the result of random patterns that form from time to time?

Absolutely not. It seems that the entire past history of our universe has been spent in a phase with incredibly well-defined and regular natural laws. Any variation in the basic physical laws would show up in our not being able to match observations with predictions beyond some key epoch when the "changeover" happened. Currently, big bang theory predictions are based on the assumed uniformity of the natural laws all the way back to something like one trillionth of a trillionth of a trillionth (10^{-36}) of a second after the big bang. No one knows what really happened before then, although there have been a few discussions in the technical literature about Chaotic Gauge Theory, which proposes that instead of a state of perfect symmetry, the universe emerged from a state of perfect chaos where there were no laws at all. It was a state of complete anarchy, with perhaps no time, space, and cause-and-effect to boot. There are many examples in nature of systems undergoing phase changes (ice to water), and perhaps initially the universe *was* in some chaotic state. Perhaps our natural laws also emerged from chaos as the universe cooled.

260 If time did not exist before the big bang, can we really say that it exists now?

This is actually a very hard question to answer. The problem is that there is no cause-and-effect link between anything that presumably happened "before" the big bang, and the conditions that arose after the big bang's expansion commenced. We can speak of time existing because, mathematically, this is one of the parameters of our world with which we can organize seemingly unrelated events. It is not a unique "dimension," but it is certainly one of a small set of necessary ones.

261 What did the dense matter soon after the big bang look like?

It was not a dense point of matter somewhere off in 3-d space because general relativity forbids such a "solution" to the gravitational field equations that are our only current guide to what was happening then. Instead, you have to think of this like a group of far-flung bacteria located inside the core of a star. Each of these observers would see essentially the same seething, dense, luminous matter in their immediate vicinity, but they would not see each other because although they are separated by a foot, light has only had enough time to travel an inch since the start of the universe! You would not know about the existence of the other bacteria until the universe had gotten old enough that light could travel that foot. Each of these bacteria astronomers would look around within their minuscule "observable universe" smaller than a golf ball and see an unbelievable sight. It would be blindingly bright because at that time the contents of the universe are dominated by the ten billion photons of light for every quark. You would be hard pressed to find any matter at all. The equivalent density of this luminous "matter" would be billions of times higher than nuclear matter but undergoing decompression as the universe expands. The temperature would be several trillion trillion degrees.

262 Are the universe and the Moon's orbit expanding at the same rate?

This question comes about because the lunar orbit is increasing in radius by about 1.5 inches per year and the expansion rate of the universe, determined by what is called the Hubble Constant, has a value of 44 miles per second per megaparsec. These numbers seem unrelated in the particular units they are expressed, but we can rewrite them into the same units. The radius of the lunar orbit is 238,000 miles. The time it takes to double the orbit radius is just 12.8 billion years if we assume that the orbit increases smoothly in time. For the universe, the Hubble Constant is a curious number that has the physical units of 1/time. When we convert megaparsecs into inches(1 trillion trillion) and 44 miles/sec into 2.8 million inches/sec, and then divide, we get a time equal to 13.8 billion years. Notice that the time it takes for the lunar orbit to double (12.8 billion) is practically the same as the time it takes for the universe to double its size (13.8 billion). Is there anything "cosmic" about this similarity? Not really, because the lunar orbit doubling time is just a reflection of the extreme age of the Moon, which to within a factor of three is nearly the same as the age of the universe. The physical reasons why the lunar orbit and the cosmos are expanding are very different and completely unrelated.

CHAPTER 7

The Cosmos— A Matter of Extreme Gravity

Gravity has been our sometimes fickle friend since the dawn of life. It helps us fall when we stumble, but it also lets us enjoy the thrill of a roller coaster ride at the beach. It has been such a commonplace element to our lives that we hardly realize its deep significance in the grand scheme of Nature. When we bother to think about it for a moment, a myriad of questions bubbles up and demands answers: What is gravity? How fast does it travel? How does it dictate the scope of space and the cadence of time? We probe its fabric with increasing clarity and see how much of what we thought before has fallen away like an old suit of clothes. What remains is a mysterious arena onto which matter and field are painted. All of our reference points that help us physically orient ourselves vanish into a patina of nuances that grace gravity's fabric like the color of burgundy wine. Exploring places in the cosmos where gravity has won the upper hand over matter reveals black holes and exotic contortions of space and time. There are places in the universe where we cannot go, and places from which we can never return. But thankfully there are more than enough space and time in our cosmos to keep us occupied and enthralled.

◐ ● ◑

263 What is the latest on black holes?

Here is a discovery that is actually pretty weird. The discovery is that the amount of information lost as material falls into a black hole appears as an increase in the surface area of the black hole. In other words, all the information you need to describe a three-dimensional object can be found in a coding of the two-dimensional surface that bounds it, even though you lose a dimension in going from the one to the other. It would be like being able to describe the state of all of your internal cells by writing information on the surface of your skin. It sounds bizarre, but this seems to have been mathematically proven for black holes. It works for the medical technology, which allows CAT scanners to study the inside of your body. It also works for the entire cosmos. Before the cosmos inflated to its enormous size, it was a vanishingly small volume of spacetime only 10^{-28} inches across. The minimum size of a spacetime volume that can contain any information at all is only 10^{-33} inches across, and so this initial volume of spacetime—the kernel out of which our cosmos emerged—contained only about 10 trillion of these quantum cells, and the surface area contained about a billion of these patches. Each patch would be inscribed with a 1 or 0, so that means you could specify all of the essential information of the cosmos on a single one-gigabyte hard drive.

264 Where can I find a list of the currently known black holes?

Tables 9 and 10 give lists of the known stellar-sized and supermassive black holes, although both are works in progress. The closest ones with masses similar to the Sun are several thousand light-years away. As for the "silent and deadly" kinds that do not produce tell-tale X rays from infalling gas, we do not think there are many because we can account for most of the gravitational field of our Milky Way without adding many of these dark stars. There are probably no more than a thousand or so stars in the entire Milky Way that will end up as black holes after they supernova, and it seems that the rate of producing these stars has been pretty constant over the last five billion years. There are probably as many supermassive black holes with masses of several million to a billion Suns as there are galaxies in the universe. It is believed by astronomers that just about any galaxy with a distinct nucleus or central core has one of these lurking there. The most active ones cause the galaxy to become a powerful radio or X-ray source, and there are thousands of these "active galaxies" that have been catalogued since the 1970's.

265 What would you experience on a trip into a black hole?

I would spend a lot of time screaming. Frankly, no one knows. I have never seen any calculations that attempt to predict in detail what the victim would actually see. The black hole models that come from general relativity usually identify different regions of spacetime only in terms of their rough geometric properties. Outside the event horizon (see Glossary), you have the normal universe we know about. Inside the event horizon, there must be enormous optical distortions that would make the scene look like something out of a carnival fun house. The radiation, depending on its source, would have various amounts of gravitational and Doppler red shift or blue shift. Beyond that, I cannot even begin to imagine. Would the singularity (see Glossary) where matter and energy reach infinite density appear utterly black or look like a luminous star in the milliseconds before you are snuffed out of existence? My guess is that it is very dark in there because I can't imagine any way for the singularity to be a source of light if nothing could escape from it in the first place. But would you see a blinding background of light from radiation flowing across the event horizon in the direction you are looking, or would the light rays make a beeline straight for the singularity and never travel to your eye? Now there's a fine homework assignment for a physics graduate student!

266   Is there a black hole at the center of the Milky Way?

Astronomers have suspected this for over a decade, and there have been many reports in the news media of "proof" that such a thing existed. Now we have the most direct proof imaginable: The motion of the stars and gas near a region called SgrA* (pronounced sadge A star) in the constellation Sagittarius. These motions are faster than what you can account for if you just add up the mass of the gas and stars you see and work out their gravitational speed. This kind of motion has now been directly detected in a star called S2 located in orbit around the black hole. An international team of astronomers led by Rainer Schödel at the Max Planck Institute for Extraterrestrial Physics observed this star over the course of ten years as it completed two-thirds of an orbit around a region centered on SgrA*. They reported their stunning results in October 2002. Figure 19 shows what the data look like, and Plate 13 is a color version of the star field, with an arrow indicating the position of the black hole. S2 approaches the central black hole to within three times the distance between the Sun and Pluto—while traveling at no less than 11 million miles per hour. The fast-moving star takes about 15 years to complete a single orbit. They used the Adaptive Optics Instrument on the 8.2-meter Very Large Telescope in Chile and captured a sequence of high-resolution images of this star as it orbited the black hole. The black hole, however, was not visible. From the parameters of the elliptical orbit of S2 around the black hole, the investigators derived an enclosed mass between two and five million solar masses. This small volume of space, and large mass, completely excludes the possibility of a massive star cluster. Only a black hole fits the data and at last settles this issue once and for all.

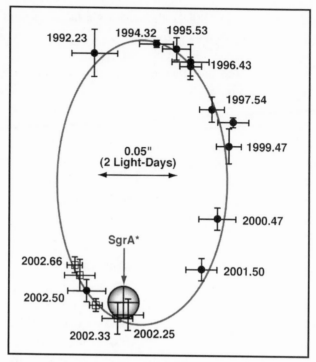

FIGURE 19 A doomed star orbits a black hole. Every 15 years, this star known only as S2 orbits a three-million-solar-mass black hole at the center of the Milky Way. This first-ever measurement confirms a decades-old proposal that a massive, though inactive, black hole hides among the bright star clouds of the galaxy's core regions. (CREDIT: RAINER SCHÖDEL, INFRARED AND SUBMILLIMETER ASTRONOMY GROUP AT THE MAX-PLANCK-INSTITUT FÜR EXTRA-TERRESTRISCHE PHYSIK)

267 Can particles get distorted when they enter the singularity?

Of course we have no way to answer this with any confidence. First of all, physicists don't have a firm understanding of the structure of elementary particles such as electrons and quarks—the basic constituents of normal matter. We also have no idea of what the singularity in a black hole really looks like at the sub-atomic scale. Does it represent a region of new physics whose rules we haven't even thought of yet? If the current ideas of string theory (see Glossary) remain sound, elementary particles are intimately related to the same kinds of string-like particles that represent gravity. So one imagines that as you approach the singularity, particles dissolve into their string-like constituents, and these somehow blend, or are transformed, into the string-like constituents of the singularity's gravitational field. Other than exploring the beauty of exotic mathematics, we will never be able to see inside a black hole to check whether this, or any other, idea really makes sense in terms of what the physical universe is actually capable of doing. Yet the demonstrated existence of black holes and of the equally mysterious cosmic dark energy during the 1990's challenges us to imagine the impossible.

268 Where does matter go after it is squeezed through a singularity?

That's something that everyone would like to know, but for now we don't have a verifiable theory of gravity (called by some people quantum gravity theory) that can take us beyond the singularity state predicted by general relativity. The singularity, a state of infinite gravitation, is a wall that we can't penetrate right now because general relativity, which is painting our description of it, cannot be tested under these conditions to confirm that it is really giving us correct answers. Physicists are convinced that this state is not real, but only a sign that general relativity has broken down. Some physicists including Stephen Hawking believe that matter entering a black hole in our universe will emerge as matter spewed forth from a so-called white hole in another universe. The mathematics seem to suggest this, but there are many difficulties in interpreting such theories without knowing whether they are accurate representations of our physical world. Our ability to experimentally test a theory has proven itself to be our only sure way of separating truth from mathematical fiction and keeping us on the right course in an astonishingly complex universe. So far, testing these new theories seems almost as hard as creating the theories in the first place.

269 Do black holes ever get full?

No, black holes never get full. The size of a black hole is fixed by the amount of mass it contains. For each increment of mass the size of the Sun, its radius grows by 1.7 miles, so if a black hole has a mass of ten times the Sun, its radius is 17 miles. The more matter that falls into it, the bigger it gets, so in fact it never fills up inside. If a black hole consumes matter at too ferocious a rate, the radiation pressure generated by the infalling matter provides tremendous resistance to the flow of matter. The rate at which matter can fall into a black hole is regulated to what is called the Eddington Accretion Rate. For a solar-mass black hole, this rate equals the mass of our Sun consumed every 100 million years.

270   Can a black hole evolve and die just as other stars do?

Actually, we think they can, but it takes a very long time. Once formed, black holes can absorb matter from their environments and continue to grow. Steven Hawking discovered in 1975 that, if left to themselves, they could even evaporate away completely. For stellar-sized black holes, this evaporation process takes 10^{100} years or more! As a comparison, the most common stars take no more than 100 billion years (that's 10^{11} years) to run through their entire evolution.

271   How can massive black holes form so soon after the big bang?

Star-sized black holes form when supernovas explode, but mid-sized (10,000 solar mass), giant (1 million), and supermassive (1 billion) black holes take longer to form because we think they require the merger or feeding of some initial black holes to start the process. Astronomers have now discovered quasars dating from a time only 800 million years after the big bang, and only 600 million years after the start of star formation. This means that supermassive black holes, or at least some of them, may require special and unusual conditions to form them. If you started with a stellar-sized black hole, you would need to feed it at a rate of up to two solar masses every year in order to grow it into a supermassive black hole. This seems like a lot to ask even of the core of colliding galaxies, especially since this feeding process is very messy. Far more than two solar masses per year would actually have to be involved. Another possibility has just been considered by astronomers who recently discovered a new class of black holes—the mid-sized holes with masses of 1,000 to 10,000 solar masses.

The Arches star cluster consists of tens of thousands of stars pack tightly together, all just 100 light-years from the galactic center, or about 25,000 light-years from Earth. Joel Bregman at the University of Michigan and his colleagues used Chandra and Hubble Space Telescope images of the Arches Cluster, and have discovered that the core of this cluster contains a massive black hole. In computer simulations by astronomer Steve McMillian, one star could collide with a neighbor, setting off a runaway growth pattern over a period of a million years. An Arches-like cluster could eventually command enough mass to form a mid-sized black hole and slowly sink toward the galactic center. Because these kinds of clusters are plentiful, one could easily imagine a steady rain of these mid-sized black holes into the core region of a galaxy. Over time, they would merge and form supermassive black holes. In reality, it may be a mixture of accretion and black-hole merger that does the trick.

272 What happens when black holes collide?

There is nothing to prevent two black holes from orbiting each other just as the Moon orbits Earth. So long as they are at a distance of a few dozen times their event horizons, they can orbit each other for a very long time. If they are much closer, then gravitational forces deform them into football-like shapes and they start to emit gravity waves in huge amounts. This causes the orbits to evolve and decay rapidly so that the black holes eventually merge together. The Chandra X-ray observatory recently discovered two supermassive black holes in the core of the galaxy NGC 6240 in the last stages of this million-year death spiral. A team of astronomers led by Stefanie Komossa at the Max Planck Institute announced this discovery in November 2002. At the present time, the two black holes are 3,000 light-years apart, but within a few hundred million years there will only exist one, even more stupendous black hole. In another galaxy called Arp 220, a similar pair of supermassive black holes have also been spotted by Chandra. Astronomer David Clements at the Imperial College London and his colleagues announced this finding in April 2002. This galaxy is well known to astronomers as one of the most powerful infrared galaxies in the universe. The bottom line is that when two black holes collide and merge, they emit huge amounts of gravitational radiation. This energy is lost to the black holes, and causes the black-hole system to lose about 5 percent of its mass in the process of formation.

273 What is the singing black hole?

Astronomers have known since the 1970's that the X-ray light produced by some black holes isn't steady in intensity. It actually flickers in intensity at many different times, from milliseconds to several seconds. Astronomers call some of the more regular of these flickerings "quasi-periodic oscillations" or QPOs. If the process producing this flickering were completely random, every frequency of flickering should be present with about the same intensity, like the hiss you hear on a blank recording tape. But this turns out not to be the case. For some black holes, slower flickerings are more common than fast flickerings, and in a particular way called "one over f" noise. What this jargon means is that whatever is going on in regions closest to a black-hole horizon, it has some kind of memory or "correlation." Astronomers Phil Uttley and Ian McHardy of the University of Southampton have analyzed this flicker noise and now conclude that it has a common cause among a vast number of different black-hole systems. They have used the NASA Rossi X-ray Timing Explorer satellite for the last six years, listening in to the X-ray sounds from a variety of black holes. What they think they are hearing is astonishing. Near the inner edge of an accretion disk, gases are turbulent and form cells of plasma. When these plasma cells of varying size pass across the black-hole horizon, they emit a burst of X rays. When added together, the X-ray light from these gas cells produce the "one over f" noise that is detected by the NASA Rossi XTE satellite. What is even stranger is that the same black hole will suddenly change its style of making this noise. This kind of noise is also interesting because if you took classical, popular, or jazz music and counted its frequency content in the same way, it would produce a similar flicker noise. So, although we may not identify the tune, black holes do have a song to sing, and, like whales in the ocean, they change their song from time to time. Even more surprising is that supermassive black holes in the cores of distant galaxies also produce this same kind of noise, but slowed down a million times. The conclusion is that stellar and supermassive black holes lead to exactly the same observations and accretion disk physics once the differences in their masses are factored out.

274 How do we know relativity actually works near a black hole?

Einstein's relativity equations seem to work just fine so long as the curvature of spacetime (the strength of the gravitational field) does not become infinite in a condition called a singularity. This does not happen near an event horizon but only happens as you approach the very center point of the black hole itself. Another way to think about this is to sneak up on the horizon from the outside and see if our predictions are still making sense. That will tell us if the equations are working just outside the horizon. One of the big discoveries made in black hole physics, and another test for the accuracy of Einstein's general relativity theory, is that just outside the horizon, objects will be dragged around the black hole by the action of "spinning space." This is called the Lenz-Thirring Effect. On November 6, 1997, astronomer Wei Cui at MIT and his collaborators used the NASA Rossi XTE satellite to uncover an unmistakable trace of spacetime being dragged around by three spinning black holes GRS 1915+105, J1655-40, and Cygnus X-1. So, this means that Einstein's mathematics does work just outside the event horizon, and so we are pretty sure it must also work a few yards away on the inside of the horizon, too. Of course, we can never really prove this because all of the possible observations we could ever make come to a full stop just outside the horizon itself. Why does a black hole have a horizon at all? It has to do with the very coordinate systems you use to describe what is going on, and these depend on where the observer is located.

In general relativity, time and space are a set of variables that can be used to keep track of the geometry of spacetime. But they are not the only kinds of variables that form a set of four coordinates that span the dimensionality of spacetime. They are just the ones we intuitively find the most convenient. Physicists have discovered other sets of coordinates that are even better. In general relativity, it is only the curvature of spacetime that tells you if something unusual is going on or not. In a black hole, the curvature becomes infinite only at the singularity. Nothing unusual happens to the curvature of spacetime at the event horizon. If you use Cartesian coordinates (x,y,z,t) like the ones in special relativity, you discover that several physical quantities such as red shift and time dilation become infinite at the event horizon. This tells physicists that, although intuitively helpful, Cartesian coordinates are not a good choice for describing physical processes near the event horizon. But if you use something like the Kruskal-Szekeres coordinates, this problem goes away. These very odd coordinates let you smoothly describe various physical quantities from infinity all the way down to the singularity itself, with no discontinuities in between. In the Kruskal-Szekeres coordinate system, which is actually the one you would be living in if you were falling into the black hole, as you pass inside the black hole event horizon, nothing unusual happens, but in the conventional Cartesian (x,y,z,t) system, you see that the space and time parts reverse themselves. This means that just inside the horizon, space becomes time-like and time becomes space-like. The particular mathematical variable we call time changes to something with the mathematical properties we have normally associated with space. It is possible that time changes, somehow, inside a black hole, but that is an experiment we will never be able to test because we can never receive information from inside a black hole. Because (x,y,z,t) is a "bad" coordinate system in the first place for describing black hole geometry, I do not feel compelled to ask it to make too many predictions.

275 Can one black hole form inside another?

No. What can happen is that two black holes can merge, and for a time measured in minutes you end up with a highly deformed, spinning object with one event horizon but a very complex inside geometry that might contain two singularities. But there is no natural mechanism whereby a black hole would suddenly form inside another.

276 Can black holes trap gravity waves?

Yes they can. Any accelerations of matter that happen inside the event horizon to produce gravity waves are trapped inside. Gravity waves escape to infinity if they are emitted just outside the horizon, but not inside.

277 What predictions of black hole theory remain untested?

No evidence has ever been found for the existence of worm holes—those extreme distortions of space that might let you travel instantly across millions of light-years. Worm holes require a specific set of conditions for them to form, and there is no known way that such carefully balanced conditions seem possible in a universe as messy as ours. Unlike black holes, they have no singularities inside them. There is no way to form singularity-free objects during the course of stellar evolution. Many science fiction authors like to use them to allow spacecraft to travel quickly from place to place in our universe. But all of these ideas are based on "pure math" descriptions of how they might work. Nature is often much messier than any ideal, abstract rendering. There are no perfectly straight lines in the universe, and there are not likely to be worm holes, either. Hawking radiation has also been predicted but never detected. Particles and anti-particles are constantly appearing and vanishing in the physical vacuum. This can be indirectly observed in numerous lab experiments here on Earth. Near a black hole, the gravitational field changes in strength due to its tidal component, and when this tidal scale equals the scale of a virtual particle pair, the pair can be tidally ripped apart. This causes one of the particles to escape to infinity and the other to fall through the horizon and be lost. The gravitational field, meanwhile, has lost energy in doing the work to separate the particles and to confer "positive energy" to the particles that escape to infinity as the Hawking radiation. The net effect is that the black hole has lost mass equal to the mass of the escaping particle. For a black hole of one solar mass, the event horizon of a black hole is located at a distance of 1.7 miles, but just outside this horizon at a distance of 2.4 miles, an incident photon is deflected into a circular "photon orbit." These orbits, however, are unstable. After a few orbits, the photon will either spiral inward into the event horizon or manage to escape to infinity with a very large loss of energy. Matter particles can orbit a black hole at distances much greater than the photon orbit radius, depending on the particle's velocity; however, for rotating black holes, the Lenz-Thirring "frame dragging" effect prevents any long-term, stable orbits for matter within about 17 miles of the black hole. Although the Lenz-Thirring effect has been confirmed by X-ray astronomers, the existence of photon orbits has not.

278 We can bring information back from the center of a hurricane— why not from a black hole?

It's not the same thing at all. The physical laws that describe these two systems have nothing to do with one another. There is no physical event horizon to the eye of a hurricane that prevents light or radio signals from reaching us. Planes that fly into the center of a hurricane can still radio back to base to say that they are there, and studying the region. All electromagnetic signals are prevented from escaping from the interior of a black hole because, according to general relativity, there is a real event horizon that surrounds every black hole, within which anything traveling below light-speed cannot escape. This horizon is a physical one, not one limited by our technological prowess. Any form of information transmitted by light-speed phenomena will be affected in exactly this way. The so-called sound barrier was never considered a physical barrier, only a technological challenge during the first half of the 20th century.

279 Why do bodies have gravitational fields?

In a deep way, we actually don't know why there are gravitational fields in the universe and how matter and energy generate them. It seems to be just a part of the way the universe and the physical world exist. There are many of these kinds of ultimate questions that have no answers, at least right now. All we can do is describe how the forces act, and their relationships to one another. That's quite a lot to do just by itself!

280 Does gravity produce gravity?

Yes it does, and this makes the mathematical treatment of this force a nightmare to work out. Gravitational fields possess energy, and energy generates gravity just the way matter does. Hence, gravitational fields self-interact. The gravitational field of Earth interacts with itself in such a way that it is as if Earth had an extra smidgen of mass beyond its known 5,976 billion billion tons. This amount caused by Earth's gravity field interacting with itself is equal in mass to about 100,000 billion tons of mass. It doesn't sound like much, and for many calculations involving satellite orbits and inter-planetary celestial mechanics, it is completely negligible. But in other systems involving orbiting neutron stars or black holes, this smidgen can actually add up to a significant percentage of the total mass of the system exceeding 10 percent. For other fields in nature, like the gluon field that holds quarks together inside protons, the situation is even more extreme. Gluon fields also interact with themselves, and they are the cause for nearly 95 percent of the actual mass of a proton. Most of the "mass" that you measure each morning on the bathroom scale is actually in the form of gluon field energy, not ordinary rest mass as you might think of it.

281 Can gravity be shielded?

No. Gravity acts on, and is generated by, both matter and energy. The very objects, fields, or containers you try to use to build a shield, fashioned from matter or energy, would produce their own gravitational fields. Gravity can only be shielded by creating some "entity" whose nature would be to block the space and time components from one region of spacetime from another region of spacetime. It is unimaginable what that entity would have to be like to accomplish this kind of blocking.

282 Can gravity be simulated using electromagnetic forces?

Not completely. Although electromagnetism can be attractive or repulsive, gravity provides only a force of attraction between all forms of matter and energy. It is possible to match the acceleration of gravity by using particles of opposite charge, but numerically all you would have is a force field that mimics one feature of gravity over a very small region of space. Take away the charges and the similarity immediately vanishes. Contrary to what some science fiction stories might imply, we know of no electromagnetic analog to gravity. We can, however, create electromagnetic force fields with charged matter that can alter the total forces they feel due to gravity. We can levitate charged particles in magnetic fields and so on.

283 Why is the gravitational field of the universe another name for spacetime?

This is how Albert Einstein chose to describe it, mathematically. And his choice led to a very successful theory called general relativity. The development of any mathematical theory of natural phenomena such as gravity requires that the mathematical symbols defining the theory must be related to qualities of the phenomenon such as the symbol T representing temperature, V representing velocity, or M representing mass. In general relativity, a similar association had to be made by Einstein. Einstein symbolically defined the gravitational field to be identical to a quantity that 19th-century mathematicians called the metric tensor. Einstein's minimalist adoption of the metric tensor as the embodiment of the gravitational field was significant and has far-reaching ramifications. Before Einstein, the metric tensor was a purely geometric quantity that expressed how to determine the distances between points in space. Einstein's appropriation of the metric tensor so that it also represented the gravitational field led to an inevitable, logical conclusion: If you took away the gravitational field, the metric would be, everywhere and for all time, equal to zero, but so too would the very nature of space itself. The distance between points in space and time would vanish. Space and time would disappear into nothingness. In the 1961 book *Relativity: The Special and General Theory,* Einstein is quoted as saying, "Space-time does not claim existence on its own, but only as a structural quality of the gravitational field." When we write our equations that depend on time and space locations, we consider this coordinate grid-work to exist in some more fundamental way than the particles, fields, and energy they are meant to locate in space and time. But Einstein firmly believed that this comfortable, intuitive view was wrong. If the metric tensor is identical to the gravitational field, which is what experimental evidence has since shown, then the coordinates of spacetime we erect to define place and time must also in some sense be constructs of the gravitational field. The experimental tests of general relativity are now so restrictive that no other interpretation than Einstein's original one survives.

How are we now to interpret the points that make up spacetime in terms of physical properties of the gravitational field? Geometrically, a point has no size at all, and space is built up from quite literally an uncountable infinitude of these points. Physically speaking, a point in spacetime is defined as an "event" that has a unique coordinate address. All observers will agree that such an event occurred, and each will assign it a unique address in their own coordinate system, but in comparing these addresses with other observers, the space and time components to the addresses will be different. An event at its most elementary level could be the collision between two particles or the emission of a photon of light by a particle. An event could be any intersection between two worldlines in spacetime. A worldline is the complete history in space of an object or particle. By filling up spacetime in this way, every mathematical point eventually finds itself near some intersection point in the net of intersecting worldlines described by the energy (light) and matter worldlines that fill up the spacetime. It is the collective property of physical events that defines physical spacetime and its geometry. The interstitial space between the events is simply not there so far as the physical world is concerned. After all, a spider is free to crawl around its web, it's just that it can't crawl around if the web is not there. Physicists generally subscribe to this idea when they think about space.

284 How does a body actually bend spacetime?

A body is not separate from the spacetime in which it is embedded any more than the color of a burgundy wine can be separated from the wine. In fact, the entire thrust of string theory and other ideas like it is to show how matter is actually built up from space. A body produces a gravitational field the same way that other bodies produce the other fundamental forces—through the action of a quantum field. In a very real sense, the field of a body is an extension of what a "body" is. It is not something that is produced or generated like a radio station generates a radio or television signal, or the way that the color of a leaf is generated by the leaf. Instead, a field is the result of a transaction of energy between objects, and for gravity we describe this transaction as the geometry of spacetime. Once we have a better quantum theory of gravity, we will probably also have a better idea how this process works.

285 Is gravity the exchange of gravitons, or the curvature of space?

Theorists believe that both of these descriptions are valid, in much the same way that we can think of the force of electromagnetism as being either the product of a continuous field, or the exchange of numerous force-carrying particles called photons. Just as photons are "packets" of the electromagnetic field, gravitons would be considered packets of the gravitational field or spacetime curvature. For certain calculations, the description of electromagnetism as a smooth field is more workable than its "quantum" description, and vice versa. You do not have to worry about the quantum graininess of gravity (if it exists at all) to calculate the path a planet takes around its star, or the way that a dropped hammer impacts your toe. Although physicists have a workable theory of gravity that involves the gravitational field, and gravitational forces as a curvature of spacetime, there is no currently accepted quantum theory of gravity involving gravitons. Physicists cannot even agree just how to describe these particles, if they exist, beyond a few very sketchy properties (for example, no charge, no mass, travel at light-speed, quantum spin assignment of 2). We don't know from the start whether there are such things as gravitons even though we have isolated the particles responsible for the other three forces in nature.

286   Is space really quantized?

Practically all physicists working on a Theory of Everything (ToE) believe very strongly that at some level, space (actually spacetime) will turn out to be quantized just like all the other fields we know about. The scale at which this may happen is nearly 100 billion billion times smaller than the nucleus of an atom—a length called the Planck length. It is impossible to measure events at this scale in normal laboratories, but according to Giovanni Amelino-Camelia at Oxford University and his international team of collaborators, there may be a way to detect the graininess of space at these scales using the entire universe as a microscope. Gamma-ray bursts produce very high-energy light, which travels billions of light-years across space. Ordinarily, light at different frequencies travels exactly at the speed of light, but if space is quantized, the higher frequencies in a gamma-ray pulse will arrive before the lower frequencies, and this delay could be measured. Their 1998 articles in the journal *Nature* predicted that the next generation of gamma-ray telescopes should be able to determine if space really is quantized the way that more Theories of Everything seem to imply (see Figure 20). There is also the possibility that we already know the answer to this question because of the very existence of black holes. Because the information contained inside the event horizon of a black hole is fixed by the surface area of the horizon (see Question 263), spacetime cannot be perfectly smooth. If it were, there would be an infinite number of spacetime volumes for which information would have to be recorded on the event horizon's surface. Since the event horizon has a finite area, the spacetime inside the black hole must contain a finite number of cells in spacetime, each the size of the Planck length. This means that spacetime must be grainy and quantized. This is considered by some physicists to be proof that spacetime really is grainy, so the entire premise of quantum gravity theory is actually verified, without anyone actually having made a direct observation of this phenomenon.

287   How does gravity travel through space?

According to general relativity, gravity takes advantage of the fact that the curvature of spacetime is mathematically defined by ten additional curvature terms. Gravitational forces are defined as the curvature of spacetime, but this

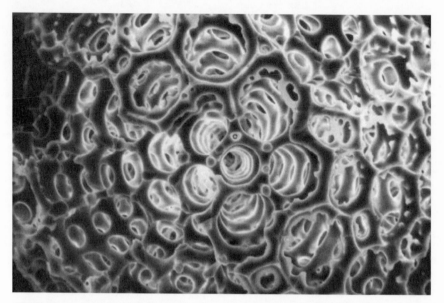

FIGURE 20 We don't know what spacetime may be like at the quantum scale, but this microphotograph of a fossil radiolarian suggests an analog to spacetime that is not so different than what string theory has in mind. (CREDIT: IRINA POPOVA AND DAVE LAZARUS, TEXAS A&M UNIVERSITY AND HUMBOLDT UNIVERSITY, BERLIN)

curvature in four-dimensional spacetimes is specified by 20 distinct terms. Only ten of these are determined by the local "source" distribution of matter. The other ten define how the spacetime *outside* a body responds to the presence of the body, and define a "source-free" solution for gravity. This only happens in spacetimes with more than four dimensions. Gravity actually doesn't exist as a force in spacetimes with dimensions of three. That is the mathematical reason why gravity can "travel" through space. There is another, perhaps simpler, way to think about this without the mathematics. In general relativity, gravity, and spacetime are exactly the same things by definition. It is impossible, within general relativity, to separate gravitational fields from the fundamental properties of spacetime. This is like trying to define what a computer is doing without mentioning its software or programming.

288   What studies have been made for gravity waves?

Lots! When physicists talk about gravity waves, they mean actual changes in the geometry of spacetime as the result of some sudden acceleration of matter. General relativity says that these disturbances will travel outward from their source at the speed of light. They can be detected by the sudden change in the geometry of space. The search for gravity waves is a very active research area that began in 1970, when Joseph Weber at the University of Maryland built the first gravity-wave telescope: a five-ton cylinder of aluminum, which was monitored for slight deformations that would signal the passage of a gravity wave. Although no such waves were ever found by Weber, this has only goaded physicists into building ever more sensitive instruments in order to detect the gravity-wave sources in the universe that astronomers are certain are there. When neutron stars or black holes spiral into each other, or when supernovas explode, huge amounts of matter are accelerated in a short time. This generates gravitational radiation just as an accelerated electric charge generates electromagnetic radiation. The trick is to build an instrument to detect these waves. These waves have the distinct property that, when they pass by, they cause objects to oscillate as the geometry of space momentarily changes in their vicinity due to tidal gravitational forces. The International Gravitational Event Collaboration (IGEC) is the first-ever network of cryogenic resonant-cylinder gravity-wave detectors. It consists of five widely spaced detectors: one in the United States (Baton Rouge), two in Italy (Legnaro and Frascati), one in Switzerland (at CERN), and one in Australia (Perth). Searching for passing gravity waves is a delicate art since it involves sensing deformations much smaller than the size of an atomic nucleus in huge detectors that are many miles in size. Gravity waves strong enough to be detected will most likely come from events such as the coalescence of black holes or neutron stars, and these are rare. IGEC reports now that in its first operational period, it has observed no gravity waves. Both the French-Italian gravity-wave interferometer, called VIRGO, and the LIGO interferometers in the United States are designed to detect very weak gravity waves by using lasers to monitor test masses placed at the ends of the arms of the interferometer. The arms in VIRGO are 1.8 miles long, while those in LIGO are 2.5 miles long. When a gravity wave passes through the detector, it causes the distance between the test masses to increase in one arm of the interferometer and decrease in the other. But not only could these gravity-wave interferometers detect black holes forming millions of light-years away, they may also be able to test various theories of quantum gravity. These theories differ in the kinds of spacetime fluctuations they predict at Planck scales of 10^{-33} cm, but the fluctuations can also affect the noise detectable in these instruments at even 10^{-18} cm scales. LISA is a constellation of three spacecraft to be launched by NASA and ESA after 2010 that uses laser interferometry to precisely measure distance changes between widely separated freely falling test masses housed in each spacecraft. LISA should observe low-frequency gravitational radiation from likely sources billions of light-years away.

289   If gravity is a distortion in spacetime, why do we call it a force?

A force is something that causes a body to change its velocity. In relativity, spacetime is the "field" that defines how strong gravitational forces will be, just as the electromagnetic field defines how strong electric and magnetic forces will be. We are embedded in the gravitational field of the cosmos just as an electron is embedded in the electromagnetic field. Changes in the strength of this field from one place to another are what we experience as a gravitational force, just as changes in an electric field cause electric forces. Fields that have the same strength from place to place in space cause no forces. For gravity, it is the change in the curvature or "warpage" of space from place to place that produces a gravitational force, but even with no curvature, the gravitational field is still present. It better be, because according to general relativity, the gravitational field is another name for the thing we call space itself. If you do away with the gravitational field of the cosmos, space and time vanish out of existence completely!

CHAPTER 8

The Invisible—
Darkness and
Cosmic Destiny

etween the familiar neighborhood of the Sun and the most remote galaxies in our visible universe, there is a vast darkness. It is the landscape of empty spaces and voids, which our best science has now filled in with a most amazing bestiary. Dark matter and dark energy are the fields that bind the cosmos and give space and time their shape. Luminous matter accounts for less than a few percent of That Which Is. In the remotest past, dark matter instructed luminous matter where to congregate to plant the seeds for galaxies and stars. In still more ancient times, even these mysterious ingredients were but a glimmer in the contents of a universe being born in fire. Space begat fields and most of these fields were "dark." A century of probing the tide pools of matter and field have led us to a magnificently more interesting and sublime cosmos than one of mere mechanical pushes and pulls. Invisible, half-real particles come and go in the void beyond human instrumentality, but not beyond human ken to experience. The very substance of our bodies is bound up in the stresses of invisible fields, not in the crude corpses of material particles. In regarding the fields and darkness surrounding us, we discover they are only another form of what we ourselves are.

◑ ● ◐

290   What is the Holographic Theory of the Cosmos?

This is a very exotic idea proposed by physicists Louis Crane (Kansas State University), Carlo Rovelli (University of Pittsburgh), Fotini Markopoulou (University of Waterloo), and Lee Smolin (Pennsylvania State University). Basically what it says is that any theory of the universe has to be one that is based on the flow of information between various parts of the universe, just as the quantum theory of matter is basically a statement about how we extract information (quantum states) from a set of observations. The universe is not a machine, it is a process or a set of transactions. This idea resembles Princeton physicist John Wheeler's participatory cosmology theory in which the observer (you) has a hand in creating the universe that it finds itself in. When you measure an electron, your experiment decides whether the electron will manifest a particle or wave aspect, but these aspects are not determined until *after* you make the measurement. It is also based on a fundamental idea about black holes discovered by Jacob Bekenstein of Hebrew University in 1973. A black hole is surrounded by a surface called the event horizon. It has been proven, mathematically, that this surface contains all of the information about the origin and contents of the black hole. The general mathematical principle that comes out of this is that all of the information within an N-dimensional object can be found on an N–1–dimensional surface surrounding that object. When we examine objects in nature, it is the surface we are examining, not the thing itself. It is a surprising, and counterintuitive, idea, but it works. When physicists apply this principle to the information in the cosmos, they uncover the idea that all of the information in our cosmos is coded in what you might think of as 1's and 0's on some mathematical surface (called the screen) that is three-dimensional, not the full four-dimensional extent of our cosmos in spacetime. The screen is exactly like a hologram. A hologram is a two-dimensional medium (film) that contains all the information needed to reconstruct the original appearance of the three-dimensional object being photographed. So, at some level, the cosmos has holographic aspects to it. The universe does not consist of things that occupy spacetime. It consists of these curious screens that we examine, and on which all of the information about the "actual" thing is coded. It is a very radical idea, but some physicists think it is one of the most elegant ways to combine what we know about black holes with the rest of the universe.

291 What exactly is a quantum fluctuation?

It is a sudden change in the energy of a particle or a field. It can also be a sudden change in the geometry of spacetime, or in the number of particles in a quantum system. They don't always obey cause-and-effect rules, and they often rely on Heisenberg's uncertainty principle to come into existence. Quantum fluctuations are the prime movers behind many of the basic atomic physics and electromagnetic phenomena in the universe and may have had a hand in creating the universe itself.

292   How is it possible for "nothing" to create a big bang?

This is one of those questions that may not ever have a sensible answer because we don't have theories that are powerful enough to describe the initial stages of the big bang itself. We can, at least in principal, speak meaningfully about the things that happened after the big bang, but it seems that we have nothing that we can test to give us guidance past that point. Even calling it "nothing" may not be correct.

293   Is vacuum genesis compatible with observation?

In a limited way. We can create electron-positron and other matter antimatter pairs out of the vacuum state by providing the vacuum state with enough energy. In a sense, these particles literally appear out of thin air, and as "virtual particles" they have measurable effects. So, yes, some limited experiments in the production of matter from the vacuum have been confirmed. Presumably, the origin of the universe is only a bigger form of the same experiment using the energy stored in the curved gravitational field of the primordial universe as the driving spring.

294   If the pre–big-bang state was timeless, what is the best guess as to what it was like?

We don't know. All we can do is watch our best current, and so-far untestable, theories predict what this could have been like. Physicists and cosmologists during the last 20 years have attempted to add quantum mechanical effects in various ways, and have come up with an initial state called the Planck era when the scale of the universe was about 10^{-33} inches at a time 10^{-43} seconds after the big bang. This is an absolute horizon to cosmology because "before" this era, all properties were determined by quantum fluctuations in some indeterminate quantum state, and these don't obey our usual temporal laws of cause and effect. There was no time or space then, not at least in any intuitive way of thinking about them. If we ever develop a true unified field theory that includes gravity, we may have more to say about what this state may have been like. But that seems to be a very far-off goal, especially insofar as actually testing such a theory is concerned. Still, speculating about this initial state is fun, and very few of us in The Profession can avoid thinking and writing about this question.

295 Will the laws of physics change in the future?

We don't know, but there have been some troubling speculations based on what may have happened long ago during the big bang. When the universe was still less that one second old, it probably underwent several freezings as the physical laws by which particles and forces act changed. We can uncover some of this change in the accelerator labs at CERN, Stanford, and Fermilab so we are pretty certain that it happened in a big way when the universe was starting out. Some cosmologists have speculated that, in the future, the universe may undergo yet another freezing as some new physical state begins to percolate and form bubbles of a new phase within the space of our current universe. By that time there will be little left of the regions of space where our laws of physics work. The current accelerated expansion of the universe exactly fits in with the theoretical idea of space making a transformation from one phase to a lower-energy one. The next round of changes may not be all that spectacular because there is not much energy left in the universe that already has a temperature only a few degrees above absolute zero. But consider this: If any of the dozen or so fundamental constants were to change by less than 1 percent from their present values, life might be made impossible. Atoms could become slightly more tightly bound together in molecules so that it would be hard to break them apart in life-sustaining chemical reactions. Also, the current cosmic energy level is only 2.7 K above absolute zero, but that "zero" could be a billion degrees above the next energy level for the vacuum. The next change might reignite a new big bang in every cubic inch of space around us today, and we would be annihilated. Don't lose any sleep over this. It could happen tomorrow, or a hundred trillion years from now. It could also happen the moment after you read this sentence. I would still buckle my seat belt and not smoke as precautions to far more likely kinds of death!

296 Would space be totally empty without virtual particles?

Since the 1950's, physicists have accepted the idea that a vacuum consists of a plenum of matter and anti-matter particles that flash in and out of existence just beyond our ability to measure and detect them. We need these virtual particles to account for how forces and matter interact. These particles violate strict conservation of energy, but we can never directly observe them as they do so. According to Einstein's definition, spacetime is the embodiment of the gravitational field of the universe itself. Gravitational fields, like presumably all physical fields, represent the interplay of virtual particles with observable forms of matter and energy. This means that if you do away with virtual particles, including the particles that produce gravity, you are doing away with spacetime itself. This is not good. Totally empty space would be empty of gravity and spacetime, and this would be just another way of saying that you have done away with the very existence of the universe as an object occupying space and time. So, eliminating invisible particles dooms our universe to non-existence!

297 What will the future of the universe be like?

For decades, students writing term papers had to repeat the debate among astronomers as to whether the universe was "open" or "closed." If open, it would expand indefinitely. If closed, it would slow down eventually and re-collapse into another fiery big bang. There was also the idea that we lived in a closed, oscillating universe that expanded and contracted in cycles lasting hundreds of billions of years. During the 1990's, all of this speculation ended. We know now that it will never re-collapse or end up in a cyclic oscillation. In fact, the expansion is accelerating thanks to a huge supply of dark energy that grows in strength as the volume of the cosmos increases. This will create a rather dismal future for us a lot sooner than the slow counting of eternity to come. In about 100 billion years, all of our galactic neighbors will have been swept away into the remotest distance of space, and the Milky Way will be alone in a dark universe. In about a trillion years, all of the stars will have "gone out" in the Milky Way with no new ones formed as the interstellar medium is eliminated and only degenerate stellar cinders like cold white dwarfs, neutron stars, and black holes are left. Beyond that era, nothing more can really be said because the timescales for anything new and interesting to happen then become of order 10^{35} to 10^{40} years when protons may decay away into electrons and neutrinos, and 10^{60} years when stellar-mass black holes evaporate by the Hawking process. Pretty bleak for life.

298 Can space exist by itself without matter or energy around?

Experiments continue to show that there is no "space" that stands apart from spacetime itself . . . no arena in which matter, energy, and gravity operate that is not affected by matter, energy, and gravity. General relativity tells us that what we call space is just another feature of the gravitational field of the universe, so space and spacetime cannot and do not exist apart from the matter and energy that create the gravitational field.

299　Have scientists proven that other universes exist?

No. There is probably no way to prove or disprove a proposition that "other universes" exist. We are forbidden from seeing beyond the horizon to our visible universe, which all cosmological models predict is only a small part of the complete spacetime that emerged from the big bang, and by definition is still a part of our universe. Still, this physical circumstance has not prevented imaginative and mathematically inclined cosmologists from peering beyond the veil and speculating what might exist there. Some cosmologists have been bold enough to suggest that in the wondrous depths of some larger spacetime, our particular big bang, and the spacetime it brought into existence, may have been only one of an infinitude of similar events, each producing its own separate universe. Many of them probably fail as universes, but others may be older than ours and have physical laws that could be very different. The problem is that these other universes are more thoroughly cut off from our own than anything we can imagine. Even if our universe is destined to expand for all eternity, it will never encounter any of these other universes. There is therefore no conceivable observation we could carry out within our own universe that could lead us to detect any of these other universes. This means that, without any hope of observational confirmation, these proposals are dangerously close to not being scientific in the purist sense of the enterprise. "If you can't test it, it ain't science!" How do you learn about a place or time that is beyond any places or times that can be reached in our universe no matter how old it gets?

300　What exactly is zero-point energy?

As with many revolutionary ideas before, it all started with a simple question: "Why in the world would the electron, finding itself in a perfectly good quantum state, decide that it had to tumble down into a lower-energy quantum state?" Even in a perfect vacuum, without a pre-existing electromagnetic field, and within a few hundredths of a microsecond, an excited electron will jump to the lowest state available to it in an atom; it is a process called "spontaneous emission." Spontaneous emission is not just some special process operating under unusual circumstances; it is the very source of the light we see from the Sun. All of the atoms in the solar photosphere experience excitation and de-excitation through the action of spontaneous emission. Why doesn't an electron just stay in its excited quantum state forever? Even before quantum mechanics were developed in the 1920's, this problem was already familiar to some physicists. Einstein and Otto Stern (1888–1969) announced in 1913 that they had discovered through their mathematical analysis that an atomic system that is emitting light seems to have a curious lowest-energy state. Instead of the energy of the absolutely lowest state being exactly zero, the system acts as though it has gotten a gift of exactly one-half of Planck's quantum ($1/2\ h = 3.3 \times 10^{-27}$ erg-seconds) from some unknown source. To make matters more complicated, the source of this energy would have to be present everywhere in space and at all times, even in otherwise perfectly empty space. This zero-point energy is present throughout space and can yield some pretty amazing effects. One of these is the so-called Casimir effect discovered by Dutch physicist Hendrik Casimir (1909–2000) in 1948. Also called the vacuum force, this phenomenon occurs whenever two conducting plates are placed face-to-face and separated by a few microns. The vacuum energy between the plates is less than it is outside them, and this difference produces a pressure on the plates that drives them together.

301 Why is the Casimir effect so intriguing?

This is a new force in nature that is caused by empty space itself. Its existence was predicted by Casimir in 1948, and it was examined in detail in 1997 by Steve Lamoreaux at the University of Washington, once physicists could construct the right kind of apparatus to measure it. If you put two conducting plates (say copper or gold) within a micron or so from each other, they will experience a force that will push the plates together. A device for measuring this force built by Umar Mohideen at the University of California, Riverside, is shown in Figure 21. This force has nothing to do with gravity or electromagnetism. It is direct property of the virtual particles that come and go in empty space but that nature forbids us from seeing directly. It is intriguing because it tells us that nature can create forces that dwell in the vacuum of space and are not produced in the same way that the familiar forces are. According to Thomas Appelquist and Alan Chodos at Yale University from work that they published back in 1983, this force may have been important in the early history of the universe by causing some of the proposed "extra dimensions" to space to become very small while others grew very large.

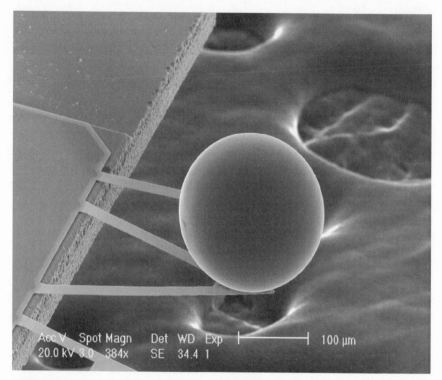

FIGURE 21 The forces of Nothingness. How do you measure the force of the vacuum? By building an instrument that is smaller than a grain of rice. This instrument, based on nanotechnology, feels a strong attractive force from the vacuum that causes the sphere to flex a sensor and register the force at the human scale. (CREDIT: UMAR MOHIDEEN, UNIVERSITY OF CALIFORNIA AT RIVERSIDE)

302 Why are Higgs fields so important to astronomy?

These fields seem to be a missing link in our most elementary understanding of matter that physicists have been looking for for decades. They are expected to provide us with clues to why things have mass, because the Higgs field seems to be the agency that gives various types of particles their mass, such as neutrinos, electrons, and quarks. They may also provide us with a clue about the nature of dark energy. Since the 1980's, the standard model of how the basic forces of nature operate has included this field as a part of space itself. Its role is to interact with electrons, quarks, and neutrinos to give them their mass, just as the color of a wine determines if it will be served with fish or beef. It also explains why the electromagnetic and weak forces can be so similar and "unified" at high energy, but so very different at low energy. The Higgs field is carried by particles called Higgs Bosons. Unlike the photon, W and Z particles, and gluons, which provide the electromagnetic, weak, and nuclear forces, the Higgs Boson carries no spin at all. Spin is a quantum property of matter that has no analogous concept at the human scale. Once you discover this spinless particle, this opens the door that there may exist other particles that have no spin as well. Theorists say that some of these spinless fields were involved in the Era of Inflation, which caused the universe to expand enormously. Cosmologists and physicists think that the mysterious dark energy causing our universe to accelerate in its expansion is related to these new spin-zero particles. If we can discover signs of the Higgs Boson in our laboratories on Earth, it may help us understand what dark energy is in the larger universe around us. So far, as Figure 22 shows, right now physicists can only use their data to estimate what the mass of this

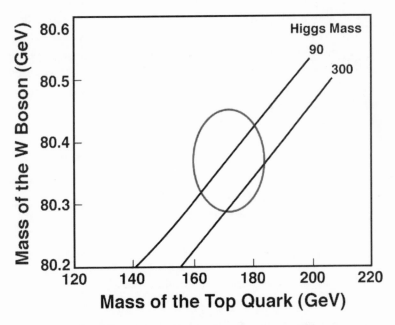

FIGURE 22 The Higgs Boson. As of 2003, this plot shows that a mass near 90 GeU is favored. The circle indicates the experimental mass ranges for the W boson and Top Quark, whose masses are believed to be dependent on the Higgs Boson's mass. (CREDIT: PETER KALMUS, QUEEN MARY AND WESTFIELD COLLEGE, LONDON, AND IOP PUBLISHING LTD.)

particle will be in today's universe. Many physicists are convinced that the next generation of experiments at CERN or Fermilab will turn up signs of this particle, which is heavier than an atom of gold—and worth far more in its discovery.

303 Can you give an example of matter created out of nothing?

You need to be careful of the word "nothing," which has a very rigorous mathematical and logical meaning but is not a useful physical concept. Instead we use the term "physical vacuum." Physicists and engineers since the early 1950's have built powerful atom smashers, which in their latest incarnations at CERN in Geneva and Fermilab in Illinois, routinely collide together particles such as electrons, protons, and their anti-particles. From their annihilation, the collisions create many new and even more massive particles, and of different types, than the ones that went into the collision. The physical vacuum is filled with half-real particles that flit in and out of existence with not quite enough energy to become real particles. The collision process allows the rapidly changing field energy of the incoming particles to provide this energy and precipitate out of the vacuum, like rain from a rain cloud, any and all of the particles that nature has deemed to be consistent with the kind of universe we live in. Because photons of light are considered fundamental particles, every time you turn on a light switch you are creating particles out of "nothing" as well.

304 Is there nothingness outside of our visible universe?

We don't think so! What exists outside of our visible universe today is probably more of the same of what we see around us right now. The limits to the space that emerged from our big bang don't appear until we reach infinity in an "open" cosmology.

305 Is there really such a thing as nothingness?

Pure mathematical nothingness, defined as the "empty set" of measure zero of all physical qualities such as dimension, energy, space, time, etc., seems to be a mathematical fiction. Nature and its quantum laws conspire to fill such a void with a patina of shifting, fluctuating quantum fields that are apparently related to the gravitational field of the universe. This field permeates our physical universe down to the so-called Planck scale. Heisenberg's uncertainty principle forbids such a state of zero energy in our universe, so nothingness isn't a meaningful physical concept.

306   Is there such a thing as hyperspace?

I don't think so, although I have enjoyed science fiction stories about it since childhood. Many theories seem to allow such alternate dimensions to spacetime beyond the four that we know and love, but when these extra dimensions are added, the observational limits to their sizes always seem to drive them into the sub-atomic domain. You can't even move a single electron or proton through them, let alone a human. Physicists have long speculated that, based on the mathematics they need to describe the forces in nature, nature must have some dimensions attached to it other than the familiar four that make up ordinary spacetime. But it is pretty well agreed that these added hyperspace dimensions are a billion billion times smaller than the nucleus of an atom, and only subatomic particles can flit in and out of them, if at all. For a particle, these extra dimensions aren't really like space at all, but something quite different in the same way that time is different from space.

307   What is a brane?

One of the unusual implications of M-theory, the current Theory of Everything that unifies gravity and everything else, is that gravity is a force that is actually 11-dimensional. Our universe, however, has only three large space dimensions, with the others (excluding time) being microscopic or even sub-atomic in size. Physicists interpret the mathematics as describing a set of three-dimensional surfaces, called branes (after "membrane") that are a part of this 11-dimensional spacetime, and that our universe is one of these three-dimensional branes, or 3-branes. According to MIT physicist Lisa Randall and Raman Sundrum at Johns Hopkins, there may be an infinite number of these 3-brane universes occupying spacetime, but separated by a microscopic distance along one of these other dimensions. We can't see these universes because none of our particles or forces, or light, can penetrate even a millimeter into these other dimensions to cross this gap. Can this theory be tested? Only the future will tell if this theory falls apart because of some as-yet undiscovered logical flaw, or if tests of gravity show that it really only works if spacetime is exactly four-dimensional. One of the tests would be to search for the force of gravity becoming suddenly stronger at atomic scales. These tests are actually in progress today, so in a few more years we may know if branes make sense. (Sorry!)

308 Does there exist a dark force in the universe?

Yes, and although we have good ideas what it may be, none of these ideas has led to the discovery of the particle or field responsible in the volumes of data available at laboratories such as CERN or Fermilab. It has no other name than "dark energy," and it is somehow embedded in space itself, as the color of a cake frosting is embedded in the frosting. It acts in a way that physicists technically call a scalar field, which means that even according to relativity, all observers in the cosmos will agree to exactly how strong this field is no matter what their state of motion. Physicists have come up with a number of theoretical ideas, each leading to a specific set of predictions about how the pressure and density of this new field should behave. When combined with astronomical observations and limits from laboratory experiments, Figure 23 gives a hint to the kinds of things that it could be, and the kinds of things that are now excluded. Still, like the Higgs Boson, this new and very common ingredient to the cosmos remains undiscovered—but hopefully not undiscoverable.

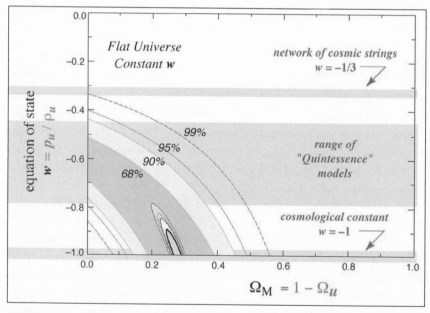

FIGURE 23 Dark energy. Over 73 percent of the cosmos is in this form of "substance," and it is causing the universe to expand at an accelerating rate. Physicists can't detect this mysterious energy directly and can only plot out its likely form by predicting what other forms of energy can do. This figure shows the relationship between pressure and density for various possible forms of this energy, and the limits currently available for them. (CREDIT: SAUL PERLMUTTER AND SNAP)

309 What is dark matter?

By measuring the speeds of stars within galaxies, and the speeds of galaxies within distant clusters of galaxies, astronomers have become convinced over the decades that there is a sub-luminous or even non-luminous something that contributes to the gravitational fields of these systems but that you cannot count up optically so that the speeds and the masses balance each other. This material can also be seen in the way that it lets hot, X-ray–emitting gases in galaxy clusters "pool" in the centers of these gravitational wells, as shown in Figure 24. This is dark matter or missing mass, and some physicists have proposed it may not even be matter. The first kind of dark matter we know about is just faint stars, distant white dwarfs, neutron stars, and black holes. These are, or once were, all made of ordinary matter (protons and

FIGURE 24 Dark matter. This image of the X-ray light from hot gas in a distant cluster of galaxies may be our only clue to the nature of dark matter. The cluster of galaxy EMSS 1358+6245 is about four billion light-years away in the constellation Draco. The mass of dark matter in the cluster is about four times that of normal matter. (CREDIT: NASA/CHANDRA OBSERVATORY)

neutrons) and have already been tallied in the weighing of the matter content of the cosmos. Next come neutrinos, which are known to carry a little mass but not enough to amount to more than a few percent of the dark matter. Finally, we have to deal with the unknown kind of dark matter that dominates the cosmos completely. We have no clue what this stuff is, only that it is probably very "cold," which means it is massive and slow moving. In the ancient universe, it was cold enough that it formed gravity wells that the hotter, ordinary, matter eventually fell into to start clumping to form galaxies. Physicists have not seen any particle that would fit this new material, but imagine several candidates called axions, gravitinos, or simply neutralinos. Until we see some traces of this stuff in our ground-based accelerator labs at Fermilab or CERN, we are pretty dead in the water in explaining what it could be.

310 How can we see through dark matter if there is so much of it around?

This is actually a big clue for astronomers and physicists that this material isn't made out of normal matter containing electrons, protons, and neutrons. These have been accounted for anyway and don't amount to more than a few percent of the universe. It is also not in the form of neutrinos, which are particles that also do not interact with light and matter very well. Physicists are busily searching for some new kind of substance that could account for dark matter and at least in their theories have turned up some interesting candidates called neutralinos, gravitinos, and axions, which would not affect light and matter but would still account for the huge reservoir we see in dark matter. Meanwhile, their traces have not as yet turned up in any of the dozens of atom-smasher labs around the world, so they must be very massive particles indeed, or they interact with matter even more weakly than neutrinos.

311 Where did the initial push that started the Big Bang come from?

Cosmologists think that the main push came during an inflationary phase when a new type of field in nature, as yet undetected, decayed. According to some theoretical descriptions, it was originally a field embedded in space itself, coexisting with the universal gravitational field, but for some reason it underwent a phase change (like water freezing to ice) that caused it to dump a lot of mass and energy into space as real particles and fields. We don't know what to call the thing that may have pre-existed the big bang. General relativity says that we cannot properly think of either a time or a place before the big bang, because these physical quantities came into existence with the big bang itself. Theoreticians, however, imagine that there may have existed some state that went through what physicists call a vacuum fluctuation. This sudden contortion of the fabric of spacetime ran away with itself to produce the big bang. Whatever started the big bang seemed to require a process of some kind to tip the right way and cause a spark of energy to flash into existence, perhaps by a quantum-tunneling event from some other unknowable state. This flash of energy was synonymous with the formation of the gravitational field of the universe, and from this fireball's expanding and cooling, the remaining particles and fields, and even time itself, had its origin. In terms of logic, we can fabricate a plausible way to get to the beginning of time and space, but along the way we find that we have no data to guide us. All the relevant details have been so badly smeared out by the inflation era that they are not recoverable in the information that is accessible to us.

312 If the big bang singularity dwarfed a black hole singularity, why did anything ever escape?

Because the big bang was a very different kind of a singularity than what you find inside a black hole. The essential physics were entirely different. Black holes are part of the local geometry of spacetime and are embedded objects like the raisins in bread dough. The big bang singularity is part of the global geometry to spacetime and is not an embedded object, so the physical conditions and processes work differently. There is an "outside" to a black hole, but there is no outside to the big bang. As we look back in time, we eventually get to see the cosmological singularity itself, something that visitors to a black hole can see only after they have fallen in.

CHAPTER 9

The Unthinkable—
Questions No One
Ever Thought to Ask

Human curiosity can spawn a bewildering supply of questions, but after a while they begin to fall into predictable patterns. A question about Pluto slowly morphs into a question about what lies beyond it. A question about events outside a black hole inches through its horizon and becomes a question about singularities and spaces within. Sometimes the best questions are those that should have been asked but never were. An astronomer can come up with many of these, but in doing so runs the risk of answering questions that no one really cares about except perhaps other astronomers! Many of the questions in this book are ones I would not have thought to ask. There are others collected in this chapter that I find interesting and had hoped someone would ask—but no one ever did.

◗ ● ◗

313   Have you ever experienced something you couldn't explain?

Yes. Thirty years ago, in a remote canyon in Yosemite Valley near Merced Lake, I was fixing dinner after a long hike, when in the distance I saw a thunderstorm approaching. It never passed overhead, but the eastern horizon was pretty cloudy. Then in the distance I heard a pair of deeply resonating, low tones about two octaves below middle-C and about a semitone apart that just seemed to come out of nowhere; they cycled between the high and low tones about four times, then vanished. I imagined that winds from the storm may have excited some kind of resonance in something, but could never figure out what. The effect was captivating, and absolutely eerie. I don't know the exact explanation for it based on science, but the mystery of this experience has lasted a lifetime.

314 Does space junk interfere with astronomy and space travel?

It certainly has in the past. The Image Science and Analysis Group at the NASA Johnson Space Center released an assessment of small-particle damage to the Hubble Space Telescope during the December 1999 mission of STS-103. The objective was to identify high-velocity impact features. They didn't have any trouble finding 571 impact features for detailed photographic study, and these had typical sizes up to 1/8-inch. If they had turned their cameras on the space shuttle, they would also have been surprised. Shuttles zoom through swarms of debris on each mission, including thousands of particles too small for tracking by the powerful Department of Defense radars. Shuttle windows, for instance, get the equivalent of a sandblasting from the debris, and NASA often has to replace its forward windows after a flight. Shuttle astronaut Frederick Hauck was the chair of a committee set up by the National Research Council to assess this space hazard. In the December 17, 1997, report "Protecting the Space Shuttle from Meteoroids and Orbital Debris," he wrote:

In 1983, three days into my first shuttle mission, I noticed a small pit in one of the windows of the crew cabin. Spectrographic analysis of the residue left in this tiny pit revealed the presence of titanium and aluminum, suggesting that the Orbiter had been hit by a chip of paint that had flaked off of some unknown spacecraft or rocket body. This was one of the first indications that orbital debris might pose a hazard to the space shuttle. By 1995, the number of reported window impacts had increased dramatically, and the debris hazard had forced planners to modify plans for shuttle mission STS-73. In 1996, significant impacts occurred on the Orbiter's payload bay door and rudder speed brake, as well as on the tethered satellite pallet. . . . The hazard from meteoroids and orbital debris is, on some missions, the single greatest threat to the shuttle and crew, slightly larger than launch risks.

Satellites have also been fatally damaged by orbital debris impacts. On July 24, 1996, the French satellite CERISE was fatally struck by the spent rocket casing from an Ariane launch in 1996. The Russian Cosmos-1275 satellite was believed to have been destroyed from a collision with debris in 1981. A quick satellite maneuver prevented the European Radar Satellite (ERS-1) from colliding with the Russian Cosmos-614 satellite in 1997.

315   Can a spacecraft be rotated to create artificial gravity?

Probably not. All you need is a toy top to see why. A spinning object in a gravitational field will soon start to precess around its spin axis. This will cause complicated forces on humans and on the spacecraft that would be physically uncomfortable, and even destructive. To provide one Earth gravity to the shuttle, it would have to spin around its long axis at a rate of 24 rpm. If you have ever ridden in a carnival ride called Vortron, you know exactly how disorienting that kind of spinning can be. You may have a sense of gravity, but the moment you try to move, your inner ear goes nuts and you feel sick to your stomach.

316   What will the Theory of Everything do for astronomy?

The Theory of Everything (ToE) is the popular name that has been given to an all-encompassing theory of matter and energy, which physicists have been working on for several decades. It is a theory that would unify the four natural forces (electromagnetism, gravity, and the strong and weak nuclear forces) into one consistent explanation for how they work, and how matter interacts with these forces. The electromagnetic and weak forces have already been unified in this way as has the strong nuclear force, which explains how quarks interact with gluons (the particles that carry the strong force), but gravity remains the stubborn holdout. The most popular theory that would qualify as a ToE is string theory, which morphed in the 1990's into something even bigger called M-Theory, but there still seem to be a lot of details that need to be worked out and tested. What impact will such a theory have on astronomy? For virtually all subjects of interest to astronomers, it will have no impact at all. Our theories of stellar evolution and planet formation, for example, will not benefit from this new knowledge. It will be most important, however, in helping us describe and discuss the events near the big bang itself, and at a time before inflation kicked in. It will also, and hopefully, tidy up physics by explaining why the constants of nature have their present values, which features of physics are derivable from others, and which ones are truly fundamental and decided by the initial conditions at the beginning of the universe. It may also launch a new revolution in thinking about the physical world if multi-dimensions and multiverse concepts survive intact and force us to confront their possibilities the way that we now accept virtual particles and dark matter.

317   How does NASA decide which astronomy programs to support?

This is a complicated process, as you can well imagine. Every ten years, the astronomical community advises the NASA Astrophysics Division about which scientific goals it considers the highest priority. The division creates a budget that includes the new missions or programs. If Congress accepts the request for new funding, or agrees with whatever diversion of resources NASA proposes in order to fund the new research programs, then the program enters the next phase. Only Congress has the full and final authority to approve new programs. NASA issues a Request for Proposal to the astronomical community for astronomers to propose what the new spacecraft mission ought to look like look like. One of those is, for example, the Mars Global Surveyor Program. The proposals are then peer-reviewed through a set of committees consisting of astronomers and other knowledgeable members of the NASA community. Research grant programs, the bread and butter of basic research, are given five to ten million dollars each year to operate. An astronomer submits a research proposal in May detailing why a particular research project is so important, why it will succeed, and exactly how much money it will cost (typically $50,000 to $100,000 for salary support). In November of the next fiscal year, NASA puts together one or more separate review committees to cover the scope of the proposals submitted to a specific program (cosmology, galactic astronomy, solar physics, etc.). NASA then specifies how much money the committee has at its disposal to fund its grants (typically a few million dollars out of a NASA budget of $15 billion). The proposal is eventually ranked by negotiations with the other reviewers, in a process that sometimes brings out the worst attitudes and biases in our community, but all behind closed doors so the proposer can never protest or know whom the committee members were. Having been a NASA reviewer, I cannot even tell you what some of these proposals were like, except in my own case. My colleagues and I submitted a proposal to use data from the 2MASS infrared sky survey to continue our study of the cosmic infrared background. We had already published two papers on our discovery of the structure in this new form of cosmic radiation, which we used 2MASS data to investigate. We proposed to use the remaining 2MASS data to complete this analysis. So far, we have been turned down for three straight years, most recently because a reviewer claimed that this work would eventually be done by satellite observatories. In fact, this is not correct, but since we have no way to challenge the reviewer's opinion, we didn't get the $250,000 we requested to finish this work, and so this entire line of research with 2MASS is now shut down. This is a major loss to astronomy, with data that the public has already paid millions of dollars to acquire and it's simply not being fully exploited.

318   Have you ever made a big discovery?

I've made some discoveries that I thought were pretty neat and that got their fair share of recognition. In the mid-1980's, for example, I discovered a dozen very weird clouds in interstellar space that looked like their shapes were being streamlined by moving very fast through some rarefied interstellar gas. One even looked for all the world like cigarette smoke! From my observations, I explored the physics of what might be going on, based on studies of aerodynamics in weakly turbulent flows, and I was happy to see that my predictions agreed pretty well with what other astronomers had also discovered about this interstellar gas. But my most exciting work happened through my collaboration with Alexander Kashlinsky in the late 1990's. We were able to use data from the COBE satellite to detect the elusive cosmic infrared background radiation. This radiation is produced by the first generations of stars and galaxies in our universe. In 2002 we continued this work with the 2MASS infrared sky survey and detected the same infrared signal in their collection of standard-star sky fields. We were elated by this result and its cosmological significance.

319   Will creating artificial life help us understand life in the universe?

It sure will. It will help us understand the chemical pathways that allow life to emerge on a planet, and to understand what the minimum requirements for a living system are so that we can identify its traces in our data. A spectacular experiment announced in 1999 by Dr. Clyde A. Hutchison, III, at the University of North Carolina and his colleagues at The Institute for Genomic Research (TIGR), found that roughly a third of the genes in the disease-causing Mycoplasma genitalium were unnecessary for the bacterium's survival. It normally has 517 genes, making it the smallest known organism, but they could get the bacterium to work with only about 265 to 350 genes. A continuation of this experiment was announced in November 2002 and is being funded at TIGR by the Department of Energy. One of the first genes they'll delete is the one that gives Mycoplasma genitalium the ability to adhere to human cells. Many of the 200 genes to be deleted make it possible for the organism to survive in a hostile environment. The end result will be a delicate creature that survives only in a laboratory petri dish. But can we create a living organism with only 100 genes? How about 50? In a separate set of experiments using compounds present in interstellar clouds, astronomer Jason Dworkin and his colleagues at the NASA Ames Research Center created an organic goo that contained small cell-like vesicles. They used a vacuum into which 100 parts of water, 50 parts methanol, 50 parts ammonia, and one part carbon monoxide were added. As ultraviolet light shined on the mixture, it was cooled to a mere 15 degrees Kelvin for two days, then allowed to warm to room temperature for another two days. This was repeated for a month and the result was a black deposit containing a number of complex molecules and small round blobs that look amazingly like little cells. Further analysis showed that these were thin walls of lipid molecules identical to cell membranes. Most living organisms squander dozens of genes to create such membranes themselves. Here you get them free of charge. Somewhere between these two experiments it is hoped we will eventually capture the step-by-step process that turns inorganic materials into self-replicating systems.

320 During which year in the next 50 will there be a number of unusual astronomical phenomena?

This question actually takes a lot of searching and comparing between numerous tables for different phenomena to offer an answer. The year 2004 actually turns up a surprising number of unusual events. The June 8 transit of Venus; the December 7 occultation of the planet Jupiter by the Moon; a blue moon on July 31, and a Christmas full moon on December 26 are not too bad to look forward to. In addition, 22 Near-Earth Asteroids (NEOs) will make close passes to Earth including Toutatis at a distance of 930,000 miles on September 29. This will also be a great year for the NASA Cassini/Huygens Saturn visit and the Messenger fly-by of Venus. Between 2018 and 2021 we have some other interesting events. A total lunar eclipse visible over North America in 2019; a spectacular close conjunction between Jupiter and Saturn (0.3 degrees) in 2020; a transit of Mercury across the Sun in 2019; blue moons on January 31, 2018, and March 31, 2018, as well as October 31, 2020; a full moon on Halloween. On October 3, 2020, the asteroid 2001 GP2 will pass within 750,000 miles of Earth. Over 190 additional NEOs will also make fly-bys of Earth within 18 million miles; a quarter of these within 9 million miles, and this just includes the list of asteroids discovered by 2003.

321 What can we do to improve science education?

We can try to remain optimistic. The biggest change has been toward more hands-on learning, which is great for exploring concepts and bonding with the basic principles of a subject. As a scientist who works with teachers through

NASA, I have had many experiences with K–12 education. By and large, it is rote memorization only, and you are tested on the mastery of literally hundreds of pieces of information every year. A recent survey by the National Science Foundation of middle-school textbooks shows that they are all useless and actually harmful to students, filled with inaccuracies and mind-deadening text. Many are even developmentally inappropriate for the intended students. Does a sixth grader really need to be tested on an understanding of subduction zones and Moh's hardness scale? More time should be spent experiencing the thrill of discovery and exploration of a few themes with immediate connection to the real world of sixth graders while a glimmer of curiosity still burns in their minds. But this places great strain on teachers, many of whom have no background in science at all. In fact, textbooks are written not so that students can use them but so that teachers who don't know science very well can learn-as-they-go from the same book the students are using. A recent 2002 National Science Foundation survey also shows that only 4 percent of elementary school teachers have degrees in science or science education. Middle school is the gateway period where most students decide to either continue in science or drop out. Young girls begin to leave science in droves, as do most minority children. Here, 33 percent of the teachers have degrees in science or science education, nearly 75 percent are women, and one in four science teachers are over 50 years old, yet nearly half of all science classes are taught by teachers who have no formal preparation in science. Those who do teach are confident about Earth and life science but not physical science, and typically space science is not covered in the depth that the average student wants, considering that space science is a very popular topic that strongly excites student interest. The U.S. Secretary for Education Rod Paige made this comment in a November 20, 2001, speech at the National Center for Education Statistics:

In 1999, the Immigration and Naturalization Service granted 115,000 H-1B visas to foreign workers. Last year, to meet the demands of our high-tech industry, Congress increased that visa cap to 195,000 workers. There is nothing wrong with the H-1B program, but there is something wrong when American schools cannot produce enough good workers for valuable American jobs. There's something wrong when foreign workers are getting jobs in America because we failed to teach American graduates the skills.

Paige was responding to statistics that show how U.S. students know less about science after high school than they did 32 years ago. Until the mass exodus of U.S students from science and technical fields in the 21st century is seen as a crisis, we will continue to rely on foreign workers. At some point, some say by 2020 to 2030, we will have to turn NASA over to a foreign labor force hired through the contracting process. What is worse is that the hiring trends seem to show that our own brand-new science PhDs are being turned away from jobs because foreign scientists are so much more mature in their careers and bring over their own money to use, but they are displacing our own young scientists struggling to get their first secure job and grant. We have indeed come a long way since the 1960's.

322   Is NASA working on a replacement for the space shuttle?

There have been several of these that have been looked into: the McDonnell-Douglas DC-X Delta Clipper, the Lockheed X-33, and the National Aero Space Plane (X-30). Considerable money has been invested in these concepts, but all were canceled without even letting the technology mature. The X-33 was killed in 2001 after a 1999 fuel tank rupture, and before any tests could be conducted on its completely high-tech approach using a revolutionary new aerospike engine design. The DC-X was a technology demonstrator built by the military and handed over to NASA; it had eight successful launches but was then canceled by NASA in 2001. The National Aero Space Plane was a single-stage-to-orbit plane that operated like a jet in the atmosphere to reach Mach 15 and then used a rocket engine to kick it into orbit. It was canceled in 1995. Currently, NASA's $4.5-billion Space Launch Initiative program hopes to reduce launch costs from $10,000 to $1,000 per pound and is in the process of selecting a new series of design concepts that involve two-stage rockets with engines fueled by kerosene, hydrogen, or a combination. One of the Space Taxi designs by Orbital Sciences Corporation in 2002 looked like a small space shuttle on top of a large rocket that can be flown back to a ground landing. The Boeing X-37 design uses lightweight composite materials and is unmanned. Unfortunately, none of these designs would be able to carry the payload weight of a single space shuttle. Instead they would function as "taxis," not "18-wheelers," which severely limits space operations after the end of the shuttle era. Even more radical designs involving Apollo-style space capsules and disposable rockets have also been considered to keep launch costs down. Those who have been raised on the sleek likes of *Star Trek* technology will be sad to see us return to the Model-T Era of rocketry to get back into space. It is hoped that the development of the winning design would start in 2006 and the first flight would happen in 2012 about the time that the current shuttle fleet might be mothballed. In the post-Columbia world, the push seems to be to disposable, low-tech 1960's style technology of the kind that the Russians keep using with great success and low cost ($2,000 per pound to low earth orbit with the Proton rocket). As you can see, there is a pattern. We fund a technology program for a few years, but as soon as we encounter a few technical problems we get cold feet and cancel the program before we have even given it the chance to mature and solve its problems. This happens at NASA all the time, and it seems hardly ever in the Department of Defense.

323   What are the two most exciting frontiers in astronomical research today?

Hands down, I think it's the search for the natures of dark matter and dark energy, and the cataloging and search for extra-solar planets. The first frontier at the cosmic scale will tell us what this "stuff" is that is controlling the destiny of the cosmos. The second frontier will eventually bring us into contact with Earth-like planets and living systems. I know of no other areas in astronomy that capture the imagination so vividly and that are as intrinsically exciting and stirring. All other topics that astronomers study have to be explained carefully to the general public, but not these. Also, these are the only two areas for which we are only beginning the journey toward understanding.

324   What is your main complaint about science fiction in the 21st century?

It has long since gone beyond the boundaries of science and has now become fantasy. Not even *Star Trek, Babylon 5,* and *Stargate SG-1* make for believable reading or TV viewing based on what we know of the current universe. For example, science fiction offers us no possible future than one that also allows interstellar travel as though it were a week-long, cross-country drive. There is not one science fiction story that plausibly describes interstellar trips taking a year, or more, except perhaps the *Alien* movies or *Lost in Space,* which use cryo-sleep to freeze humans so that they sleep for years or decades. Beyond this, most science fiction movies or TV series even reject the simple courtesy of naming their fantasy stars after ones we know about in detail. Not since Robert Heinlein's (1907–1988) *Orphans of the Sky,* published in 1963, has an author thought about long-duration space travel. Not one story in my recent memory offers a future where we are stuck in this solar system and are challenged to make the best of it. Not one story is consistent with general relativity or even Newton's laws of motion. Science fiction authors spend too much time aping each other's baseline technology and don't realize that the technology makes a civilization and the people who are a part of it. They are just as stuck with their fantasy technology (warp drive, time travel, etc.) as pure fantasy writers are with magic wands and potions. I think it's time for some new ideas and new blood. I find it very hard to believe that writers can't write exciting and challenging stories that don't rely on warp drive or worm holes. Let's get real.

325 What will be our ultimate limit in seeing distant objects clearly?

The world of high-resolution astronomy is a mathematical world bounded by distance, size, and angular scale. The human eye and most single-aperture telescopes crowd the lower reaches of this mathematical universe. Radio and optical interferometers now in use have begun to explore the vast middle ground extending to 0.0001 arcseconds. The high frontier, however, is vacant and portends a staggering view of the universe we can only dimly comprehend today. Consider just a few of the possibilities. At one micro arcsecond, the disks of planets like Jupiter would be resolvable as far away as 3,000 light-years. Earth-sized planets orbiting nearby stars would be discerned as clearly as we see Jupiter with a six-inch telescope. The motions of distant galaxies would be detectable in a single human lifetime, as would the motions of individual stars in nearby galaxies. At one nano arcsecond, the surfaces of planets orbiting Alpha Centauri are seen as clearly as Magellan now shows us Venus. We begin to resolve planets like Jupiter within the Andromeda galaxy and the surface markings on individual red giant stars out to the Virgo Cluster. The motions of distant quasars would be discerned within a few years. Beyond this limit the mind reels at even more stupendous possibilities. Unfortunately, the laws of physics present us with three formidable obstacles to overcome in making these possibilities a scientific reality: High resolution necessarily means having to build large costly instruments; not all surface details are bright enough to be seen at every conceivable wavelength, distance, and resolution. At some point you run out of photons. Because everything in the universe is motion, we have only a limited amount of time to form a sharp image before the objects move and cause image blurring. Compliments of nature's quantum limitations on instrument size, background noise, and surface brightness, most of the regions in our mathematical universe of resolution appear as unreachable as the inside of a black hole. Even so, there is still much that could be done. To a distance of 3,000 light-years there would be over a billion stars to investigate, and potentially as many solar systems. Continents and weather systems on the planets orbiting the thousand or so nearby stars could under favorable circumstances be mapped and monitored for changes. Long before the first space probes are launched out of our solar system to discover Earth-like planets around other stars, astronomers will already have identified all such worlds out to hundreds of light-years using one micro arc second imaging technology, and at orders of magnitude lower costs.

326   Are there sky events that should have been noticed but weren't?

The most dramatic celestial events we know about are the supernovas. In 185 A.D., there was only a single entry by Chinese observers for this dazzling star in the constellation Centaurus. G135.4-2.3 (RCW 86) is the likely remnant of this explosion. A second supernova in 393 A.D. was also mentioned by Chinese observers within the curve of the tail of Scorpius, now identified with the young supernova remnant G11.2-03. No other mentions of these events were found in Mediterranean or European records. The Crab Nebula supernova appeared in 1054 A.D. but was only mentioned by Chinese and Japanese observers even though its brilliance exceeded that of Venus for many months. The 1181 A.D. supernova in Cassiopeia, which formed the radio source 3C58, was also noted by Chinese and Japanese observers only. Were any supernovas seen before the 185 A.D. event? Chinese astrologers were very close observers of the heavens since 2000 B.C., but there are no records of any new "guest stars" being seen, possibly because the records are too fragmentary, or astrologers didn't pay them much attention compared to eclipses and sunspots. Meanwhile, no one has explained why European and Mediterranean civilizations between 185 and 1181 A.D. gave no indication of interest or awareness in three supernovas.

327   How do lightning storms produce X rays and gamma rays?

We don't really know, which is rather sobering because for a long time scientists thought they more or less understood how lightning works. This wake-up call happened in two steps. Terrestrial gamma-ray flashes (TGFs) were first detected by the Compton Gamma Ray Observatory in 1991 and were eventually traced back to large, severe lightning storms going on below the satellite. They last only a few milliseconds, and mimic the true celestial gamma-ray sources in many ways. By some means, severe lightning storms can generate gamma rays that shoot out of the top of the storm clouds and escape into space. But it isn't the lightning that causes these flashes—they simply do not carry enough energy. Instead, the gamma-ray flashes are being linked to powerful electric fields between the thunder cloud and the base of the ionosphere, which produce phenomena called sprites and elves, but no TGF has ever been seen to coincide with a sprite discharge. Then in January 2003, Joseph Dwyer of the Florida Institute of Technology and Martin Uman of the University of Florida announced that they had detected both X rays and gamma rays just before a bolt of lightning hurled down thin, rocket-born wires that had been designed to attract such bolts. These traces were recorded in nearly all of the lightning strikes and suggested that an 80-year-old idea that lightning creates relativistic electrons may have been right after all. For astronomers, any terrestrial process that generates X rays and gamma rays is intensely interesting because so many of our most intriguing astronomical objects are also powerful sources of these radiations. If nature has some new way of generating them, we need to be able to add it to our tool kit for studying the rest of the cosmos. As of 2003, the mystery remains to the origins of both the X rays and gamma rays.

328 Why do bones become brittle in space?

During space missions, astronauts suffer about ten times faster bone and muscle loss and deterioration than patients with muscular dystrophy and osteoporosis. For every month in orbit, astronauts lose about 2 percent of their bone mass. In comparison, a loss of 4 percent a year is considered severe for patients with osteoporosis. The molecular basis for this loss is not yet understood, and that's why so many experiments are performed on astronauts in space to gather data and help us find out. One recent, and very exciting, breakthrough was announced by Clinton Rubin at the State University of New York in Stony Brook. By subjecting mice to ten-minute daily shakings at 90 times every second, they could stress the bones enough to make the cells think they were being overworked, causing them to grow new bone tissue. Because weight-bearing activity promotes bone growth, astronauts must exercise in space to slow down bone deterioration in what has now become a standard, though very time-consuming, part of their daily regimen, but crew members using exercise as a countermeasure still lost an average of 4 percent of bone mass over the 84-day mission. During the 237-day Soviet Soyuz T-10 mission, the cosmonauts lost bone in spite of up to four hours of daily exercise. What is even more troubling is that the bone loss suffered in flight is not fully recovered after return to gravity. Experiments performed on bone cells flown in the space shuttle have pretty well pinpointed that gene expression, cytoskeleton, and nuclear structure are all changed when compared to ground controls and to onboard 1-g controls. No amount of exercise can cure this problem. What is needed is a cellular-level cure carried by a vaccine or other treatment that will temporarily short-circuit the genetic machinery involved in actively removing bone calcium. That is a tall order, and it's no wonder space medicine specialists have been working on this problem day and night for about the same amount of time as other doctors have been searching for a cure for cancer. This remains the biggest challenge to travel to Mars, or even establishing a lunar colony. Humans will *not* travel beyond the Moon unless we can solve this major medical problem. It will not matter if we have a dirt-cheap way to get to Mars beforehand.

329   What will we discover in the next five years?

Many wonderful things! In 2004, the Cassini spacecraft will arrive at Saturn. Its Huygens probe will descend through the atmosphere of Titan and tell us many new things about that mysterious satellite. The MESSENGER mission will arrive at Mercury and complete its extensive mapping mission. The BeppoColombo mission from ESA will drop a lander onto Mercury and give us surface and seismographic measurements. Mars will have been studied by a new generation of landers and rovers to uncover what the below-ground landscape looks like and perhaps find actual traces of water. The NASA Stardust comet sampler will have returned to Earth a sample of an actual comet, and its penetrator will have made a permanent crater on comet Wild 2 to examine its surface rigidity and composition. There will probably be nearly 100 additional planets beyond our solar system to marvel at, and perhaps one or two more transits from which to get the atmospheric chemistry of these distant worlds. The SIRTIF infrared telescope will have been in operation for three years and returned detailed images of the first generations of stars in the universe. Astronomers have only recently been able to investigate astronomical objects the way they appear at all of the important electromagnetic wavelengths. Plate 14 shows what such a composite image looks like, and how much more information can be extracted from the "images" of astronomical objects at other wavelengths. In a different, but critical, direction of research, in five years or so, the high-energy physics community will have a new accelerator online, the LEP II in CERN, which will continue the search for the Higgs Boson and extra-dimensions. The Planck mission, a follow-up to WMAP, will probe the cosmic microwave background at even higher resolution and reveal more details about the big bang.

330   Where do cosmic rays come from?

Cosmic rays are fast-moving electrons, or the nuclei of elements, which travel through space. They are not a form of electromagnetic energy or neutrinos. Cosmic rays were first detected in 1912 when Austrian physicist Victor Hess (1883–1964) took a sensitive electroscope on a balloon ride, and noticed that the device charged up at high altitudes. He won the 1936 Nobel Prize for this discovery. Since then, cosmic rays have been the subject of intense study to track down their composition, energy, and directions. Although some are produced by the Sun at low energy, we think that the more common cosmic rays with energies below ten billion billion volts come from sources inside the Milky Way such as pulsars, supernova remnants, and black-hole accretion disks. Those with energies above this limit must come from outside the Milky Way, with a strong concentration toward the Virgo Galaxy Cluster and the active galaxy Messier-87 that has a powerful jet of plasma leaving its core where a supermassive black hole is busy at work. Other, low-energy cosmic rays called anomalous cosmic rays are produced when the solar wind encounters the interstellar medium far beyond the orbit of Pluto. The particles are apparently accelerated by the encounter and some are returned back to the inner solar system, though with much more energy.

331   Has SETI discovered anything yet?

There have been a dozen close calls since the Search for Extra Terrestrial Intelligence (SETI) began in the late 1970's. On September 15, 1998, the Project Phoenix SETI at the giant Arecibo Telescope in Puerto Rico was studying a nondescript M-star called EQ Pegasi about 22 light-years away. It was a narrow-band rapidly moving signal in the direction of this star, which later turned out to be a man-made satellite. In 2001, scientists from the University of California's Lick Observatory, the SETI Institute (Mountain View, California), U.C. Santa Cruz, and U.C. Berkeley attached the Lick Observatory's 40-inch Nickel Telescope to a new optical pulse-detection system capable of finding laser beacons from civilizations many light-years distant. Unlike other optical SETI searches, this new experiment is largely immune to false alarms that slow the reconnaissance of target stars. Meanwhile, an $11.5-million grant from philanthropist Paul Allen has ensured that the SETI Allen Telescope Array will be built in California. Currently, three radio dishes have been built and another 347 are on order: Small and inexpensively built 20-foot dishes, they will be combined into a single "dish" covering 10,00 square yards of land. The array will be operational in 2005 and will push the SETI search far deeper into space than previously possible, without other astronomers grabbing the time from them to do "more important" research. One is reminded of the 1996 movie *The Arrival,* where the fired radio astronomer played by Charlie Sheen commandeered hundreds of rooftop satellite TV dishes to create his own SETI array. Wouldn't it be grand if Bill Gates would donate $1 billion to basic non-NASA astronomical research? Imagine what we might discover!

332 Does anti-matter fall faster than normal matter?

We don't really know. It is actually a very difficult experiment to conduct. At the scale of the atom and sub-atomic particles, gravity is so weak it hardly affects anything. Near Earth, bodies fall at 32 feet/sec/sec vertically. Anti-protons and anti-electrons when they are produced in "atom smashers" usually live only 100 millionths of a second. During this time, gravity would cause them to "fall" or "rise" by about the diameter of an atomic nucleus! It is an important experiment, however, and physicists are working on experiments that hope to test this very important prediction by Einstein's general relativity. The neutrinos detected from Supernova 1987A included anti-neutrinos, yet these particles of anti-matter arrived at the same time as the matter neutrinos. This observation shows that the two populations of neutrinos fell through the same gravitational fields and arrived at the same time. This implies that at least matter and anti-matter neutrinos fall at the same rate to one part in about one million. It is expected that within the next five to ten years, the necessary experiments with artificial anti-matter hydrogen atoms will settle this issue.

333 What is the most energetic form of light that astronomers have ever seen?

These are the so-called gamma rays. Astronomers and physicists measure the energies of these photons of light in terms of a unit called the MeV. One MeV equals one million electron volts, and this is about one million times more energetic than the photons of light we see with. A single gamma-ray photon with one MeV carries an energy equal to 0.0000016 ergs, which is an almost measurable pulse of energy by conventional laboratory standards. The most energetic gamma rays are studied by using Earth's atmosphere as a detector of what are called "air showers." When such a photon collides with an atom in the atmosphere, it unleashes a cascade of nuclear particles, electrons, and anti-electrons that reach down to the ground in a triangular cone. As seen from the ground, this event would appear as a very faint burst of light covering a large fraction of the sky. Special telescopes have been built to search for these faint pulses. One of these is the so-called Flys Eye telescope operated by the University of Utah since 1981. The telescope is sensitive to gamma rays with an energy of 1,000 trillion electron volts (that's 1,000 million MeVs). It is hard to imagine a single photon of light carrying 1,600 ergs of energy—nearly as much energy as a mosquito in flight. Many of these events have been seen over the last few decades since people began looking for them in the sky. But they arrive about one air shower each second. Meanwhile, the South Pole Air Shower Experiment ran for ten years and was finally shut down in 1998 because it never detected a single gamma-ray photon with an energy higher than 100 trillion volts. There is clearly a lot we don't know about these energetic forms of light in the universe and what bizarre phenomenon produces them.

334 Why is space shuttle computer technology so primitive compared to what I can get at my local store?

Because it has to be ruthlessly reliable in the radiation-laced environment of space. High-energy particles can penetrate sensitive circuitry and cause computer glitches that could alter critical calculations, or even cause flight computer crashes. To make computers work well in space, their circuits have to be designed to be larger than what you would find in the integrated circuits in store-bought computers. This means that the computers will be slower in performance. They also have to run software that constantly checks for faulty or erroneous commands. A February 24, 2003 article in the *Washington Post* following the destruction of the Columbia Space Shuttle seemed to bemoan the fact that the shuttle had only been upgraded once in 1988–1989. Only recently had NASA even begun to work with 1994-style Pentium processors. This made it sound like NASA was intentionally using antiquated technology to operate its 30-year-old Shuttle fleet. But if you read the full article, it correctly noted that this older technology is far more reliable than modern off-the-shelf computer designs because astronauts cannot tolerate even one computer crash in orbit, unlike the typical ground-based computer user who can merely reboot and carry on.

335 What is all this I have read about the "end of science"?

The basic idea is that some authors feel we are not making any progress in answering the "big questions." Some of the popular theories, such as string theory and inflation, are considered by some to be borderline science: They promise a sweeping new understanding of the physical world, but they are not testable with any conceivable technology we are ever likely to develop. These theories are, therefore, technically falsifiable, but in practice are not. They are also developed through a plea of "beauty" in the eyes of theoreticians who in very small numbers are the only ones able to "see" the mathematical beauty. The rest of us scientists not fluent in the mathematics are doomed to only see the beauty through second-hand translators. There are several questions that define the foundations of science that we are not much closer to answering than we were a century ago, such as "Does life exist elsewhere?", "How does consciousness emerge from a brain?", and "What is the nature of matter?" I think this reading on the progress of science is rather pessimistic since it considers the development of relativity, quantum theory, and organic evolution as some type of window dressing that doesn't really get us closer to answering the Big Questions about life, matter, and the universe. Some authors also seem concerned that our theoretical revolutions are increasingly of the kind that pits one "beautiful and untestable theory" against another so that in the end we may be left with a fragmented scientific world where in some sense you are free to choose the kind of answers to the big questions that you like, without much risk of contradiction. My own impression from the ranks of the working scientist is that "we" find these kinds of summaries rather unimaginative. It is all too easy to draw a line and say, "Science will never advance beyond this point." In fact, I find myself falling into that way of thinking about once a week, and usually on Friday afternoon. I think that at some point we will recognize that we are capable of asking zillions of questions that can never be answered, and that we will eventually decide that our understanding is good enough for now. This will be hastened when society withdraws funding for the necessary research in view of other goals it wants to commit to. Meanwhile, what's the hurry? If it takes another 200 years to answer one of these questions, I think that's great. We need to leave something for the next generation! I find it pretty arrogant to propose that just because the current generations cannot see how to answer these perfectly reasonable questions, we have to preordain the next 100 generations to believe with us that the search is pointless! Those who say that science has a boundary don't understand what science is all about no matter how learned their prose. Even though the Superconducting Super Collider was "killed" by Congress in what seemed like a stinging defeat for scientific research, perhaps that is just as well. In another 30 years, engineers will develop ways of accelerating particles that will be 1,000 times more efficient than the dinosaur technologies we use today. The stimulus for this will be the realization that less expensive means must be found or society will not support the effort.

CHAPTER 10

Nuts and Bolts—
Space Travel and
Elementary Physics

Some questions seem so basic they can't be easily placed in one of the common astronomical categories. Nevertheless, they seem intimately related to space and our exploration of the cosmos. They are about not stars but "star drives." They are about not the limits to space but the limits to knowledge or how we try to assimilate it. While questions about the Star of Bethlehem seem appropriate for a category such as Stars, a question about the religious perspectives of cosmology seems a better candidate for this chapter. The human exploration of space is a strong theme, especially with the tragedy of the Columbia Space Shuttle, which occurred during the writing of this book. Again we find ourselves questioning the wisdom of manned space exploration and underestimating its dangers. Although the questions and answers were largely written before 2003, some of the questions and answers seem entirely appropriate today.

◖ ● ◗

336   What is all the funny business about the speeds of distant space-craft and relativity?

In 1998, John Anderson at the Jet Propulsion Laboratory made the rather startling announcement that the Pioneers 10 and 11 and Ulysses spacecraft were not obeying the expected laws of gravity formulated by either Newton or Einstein. Both spacecraft seem to be falling behind their schedules for arriving at specific distances from the Sun, and they acted as though there were some mysterious retarding force pointed in the same direction as the Sun. This would be also interpreted as a change in the "inverse-square" force law at large distances from the Sun. Called an "anomalous acceleration," it amounts to about 0.0000001 cm/sec/sec. Many different effects have been looked at, including solar wind interactions, photon pressure from the Sun, outgassing of material from the spacecraft, and corrections to Newtonian gravity from general relativity. None of these seem to be enough, or in the right directions, to cause a steady deceleration of these three spacecraft. According to Anderson, the Voyager spacecraft cannot be used to verify this effect apparently because its propulsion systems and orientation methods differ from those of the Pioneers and Ulysses. Philip Laing, working under a NASA grant at the Astrodynamics Science Co. in Los Angeles, was also able to eliminate various dark-matter ideas, thermal spacecraft emission, errors in planetary ephemeris predictions, and many other rather subtle non-gravitational causes. The most promising explanation that remains viable today that doesn't require revising our understanding of gravity is the rather boring prospect of gas leaks in worn-out valves. Unfortunately, this can't be tested very easily since the spacecraft cannot be recovered. So, what do we do now?

337 Why doesn't the space shuttle burn up when it re-enters the atmosphere?

Because its underside is covered with over 20,000 individually machined ceramic tiles that have a melting point of over 5,000 F. They are so poor at conducting heat that their outside surface can remain cold to the touch while their back surface a few inches away can be heated to over 1,000 F for several minutes. Since re-entry takes only a few minutes, the tiles can get rid of the heat they absorb from the re-entry friction without, apparently, much of it building up on the side nearest the shuttle's airframe. Some of the tiles receiving the greatest heat load may also partially melt. The tiles have to be replaced by fresh ones after each trip. Unfortunately, this technology has its limits, and these were well known at the dawn of the shuttle's design. If a tile is damaged in space by, say, being hit by the ever-present space debris, you cannot replace or repair it. Each tile is unique, and must blend completely into the curve of its surface, or re-entry aerodynamics will drastically change. Astronauts do not have repair kits for these tiles because no adhesive can be applied in the vacuum of space without actually making the situation worse and risking a potentially fatal unzipping of the tiles from the surrounding areas.

338 Is NASA working on new propulsion technology to replace rockets?

NASA has been working on ion-propulsion and nuclear-propulsion technologies for decades, and ion thrusters are commonly used on satellites to provide gentle station-keeping nudges to keep them in the right orbit. On

Saturday, October 24, 1998, NASA launched the Deep Space-1 spacecraft to perform an asteroid fly-by later in 1999. The DS-1 ion motor uses xenon atoms as the fuel by first ionizing them inside the rocket motor, then accelerating them to over 70,000 miles per hour by using an electrostatic voltage potential of a few kilovolts. The motor operates day after day and produces an acceleration equal to about 0.02 pounds of thrust. This is about the weight of a piece of paper on your hand. After a year, this leads to a change of velocity of nearly 8,000 miles per hour, which is enough to allow interplanetary journeys out to Jupiter. On January 31, 2003, an enthusiastic President Bush announced his support for a true nuclear propulsion program called Prometheus. This $3-billion, five-year program will culminate in the launch of the Jupiter Icy Moon Orbiter early in the next decade. The nuclear reactor will be rated at 100 kilowatts and provide nuclear-electric propulsion along the same lines as the ion rocket motors that have flown on the DS-1 and upcoming ESA BepiColombo Mercury mission. The travel time will no longer be limited by what thrust we can impart to the spacecraft at Earth, or through ingenious loop-de-loops and gravity assists among the planets, but by the direct application of constant thrust for month upon month. This is the beauty of ion engines, so long as you can continue to supply them with electricity.

Is something even more exotic on the drawing boards? Of course! Solar sail technology is considered near-term, but the first privately funded launch by a Russian ICBM on July 24, 2001, led to failure when the payload did not disengage from the re-converted rocket booster. Another propulsion design has a spacecraft unfurl a magnetic field containing a trapped plasma. As the solar wind collides with this plasma, the spacecraft is dragged along at speeds that could approach one million miles per hour. There is even the exciting new idea of traveling by gravity assist using the complex Lagrangian surfaces of our solar system. It is a slower style of travel, but it costs no fuel at all, like bottles cast adrift on ocean currents. And what about warp drive? The universe that we have to work with doesn't seem to follow the rules that decades of science fiction story writing insists it does. Physicists are the true experts in pushing things to near-light speed, and there has never been found a trace of an entry into the super-luminal world. It would be easier to move a star than to create stable worm holes that would not shred a human into wriggling quantum spaghetti!

339   How much money do you make?

I look at the Sunday newspaper *Parade Magazine* every six months when they publish their "What People Make" article. I never see any scientists included among the hundreds of professions that are interviewed. Don't you think that's a bit weird? So, here is the scoop on one small profession among the thousands of scientific careers that are possible. After living on a graduate student salary of $9,000 each year for seven years, my first real job as a professional astronomer was exciting. When I had just gotten my first job in astronomy at the Naval Research Lab's Space Sciences Division in Washington D.C., back in 1982, I brought home $36,000. I recall asking myself the first month, "Geez. How are we going to spend all this money?" Flash forward to 2002. My take-home pay last year was $90,997, which serves to support my small suburban family of four in the Washington, D.C., area in a comfortable fashion, at a standard of living similar to other professionals of my age (50 years) in this region. My family and I have no problem spending this money. So, the next time you read those *Parade* surveys, keep a career in astronomy in mind, too. We rock!

340 Who invented the light-year and the parsec?

If I look through my antique astronomy and physical science textbooks from the 19th century, such as the 1844 *A System of Natural Philosophy* by J. L. Comstock, the section on stars cites their distances in terms of miles, and that the nearest "fixed" stars are so distant that "light cannot pass through [the space from the star to Earth] in less than three years." By 1873, *Fourteen Weeks in Descriptive Astronomy* by J. Dorman Steele still uses miles as a unit of distance, and the author notes that "The average time required for the light of the smallest stars which are visible to the naked eye to reach Earth is about 125 years." Astronomical papers between 1870 and 1890 use parallaxes measured in arc seconds but no specific naming of parsec as a shortened way to say this was in use by professional astronomers. Richard Allen's 1899 book *Star Names: Their Lore and Meaning* uses the term "light-year" but some of the older journals since 1870 also used this term. Francis Roger in *5000 Years of Stargazing* (1964) mentions that William Herschel may have used such a unit for his mapping of the universe. The French astronomer and popularizer Nicolas Flammarion (1842–1925) mentions it in an 1880 book *Popular Astronomy*. So somewhere between say 1780 and 1880, "light-year" had been used, but not at a level where it was common to describe star distances in this way. Astronomer Herbert Turner is usually credited with proposing "parsec" in a paper about stars in the constellation Monoceros published in January 1921; in 1913 Astronomer Royale F. W. Dyson wrote, "There is a need for a name for this unit of distance. Mr. Charlier has suggested Siriometer, but the word Astron might also be adopted. Professor

Turner suggests parsec, which may be taken as an abbreviated form of 'distance corresponding to a parallax of one arcsecond.'" So by at least 1913, "parsec" had been used informally.

341 Will we ever go back to the Moon?

Some people seem to think so. Personally, I'm not so optimistic, at least within the next 30 to 50 years. There are just too many other economic issues that seem to require attention. The lunar and Martian surfaces are so different that I don't think anything would be gained by going to the Moon first. The 1960's were a unique time in American history. Our economy was expanding greatly, we had no national debt, and the various entitlement programs had not yet begun to compete with other expenses in our society. This is not the case today. After a brief eight-year era of no deficits and great wealth during the Clinton Administration, we are now back to huge national debts during the Bush Administration. I remember how disappointed I was when President Nixon and Congress curtailed our lunar exploration, just at a time when we were a few years away from setting up a permanent colony there. The 1990's could have been very exciting time for us. Knowing that there were people actually living on the Moon could have inspired a whole new generation of students to pursue science and technology. But when you reduce the Space Program from 5 percent of the federal budget to 0.7 percent, don't expect us to get anywhere soon—or safely. All we have is a fleet of aging space trucks that take us to Earth orbit, and to an expensive space station that no longer serves its original purpose thanks to budget cutbacks and overruns. But worst of all, we have no inspiring destination for our astronauts and scientists to go to once they get into space. It is hard to get students to think seriously about these subjects as a career choice to really work toward, when even the manned space program can't marshal the support it needs to really do its job well.

342 Will NASA ever revisit Mercury?

Yes. In fact the space probe is called the Mercury Surface, Space Environment, Geochemistry and Ranging MESSENGER satellite. It will be launched in March 2004 and after a few fly-bys of Venus (June 24, 2004, and March 16, 2006) will enter into a stable orbit around Mercury about April 6, 2009, after completing two fly-bys of that planet (July 21, 2007, and April 11, 2008). The repeated gravity assists from Venus and Mercury are needed in order for the probe to match velocity with the very fast-moving Mercury, which travels at a speed of 115,000 miles per hour compared to Earth's leisurely 69,000 miles per hour. Because of the wacky laws of gravity physics, you actually have to speed up in order to fall closer to the Sun. You can't get there by slowing down and "falling" inwards! The $286 million spacecraft, weighing about 100 pounds, will come within 80 miles of the planet's surface and be able to image features on the surface during its planned 12-month reconnaissance. Just as MESSENGER arrives at Mercury, the European Space Agency (ESA) will launch the BepiColombo Mercury probe on an Ariane 5 rocket, which will be twice the size of MESSENGER and arrive at Mercury in only 2.5 years. It will use a continuous-thrust xenon-ion engine. Many of its sensors are similar to those of MESSENGER, but in addition it carries a pair of surface lander/penetrators, which will allow the first such landing on that planet by human technology. They will be dropped onto the north pole of the planet, set up a seismograph station, and examine the surface for traces of water ice.

It has been 30 years since Mariner 10 first flew by Mercury on March 29, 1974, and snapped a few thousand images, which many of us were stunned to see. Some now grace every textbook on astronomy you have ever seen since then. We still know rather little about Mercury. It has a very thin atmosphere of oxygen, helium, and argon and an Earth-like surface magnetic field 1/130th as strong as ours. With no significant barrier to prevent it, the solar wind pushes down to the surface, which is alternately baked at 850 F and frozen in nighttime temps of −274 F. Its core of iron occupies nearly 75 percent of the volume of this planet. Since formation, it has shrunk two miles, carrying the crust with it and causing many curious surface markings. It was also hit by a massive object that created the 800-mile diameter Caloris Basin and a large region of disturbed terrain opposite the basin as the seismic waves were focused through the planet's interior. The bar-magnet shape of its magnetic field suggests that Mercury's core is still partly molten.

343 How long will it take to get to each of the planets by various means?

It depends a lot on the particular trajectory that you take. Usually, the trajectories are in the form of a "great arc" that gracefully connects a launch time at Earth with a destination point. These arcs are usually many times longer than the straight-line distance between the two planets at a particular moment in time. For reduced-cost travel, astrodynamicists often rely on gravity-assist slingshot orbits from the inner planets to reach Jupiter, and from

Jupiter to reach more distant worlds. These loop-de-loops add years of extra travel time to a mission. Let's assume for our calculations that we just take the simplest direct approach and use the minimum opposition distance between Earth and a planet. Table 11 gives you some sense for how long it takes to get to each planet at different speeds. The space shuttle, of course, can't leave Earth orbit, but its speed is typical of manned spacecraft. The Galileo spacecraft, which recently explored Jupiter, traveled twice as fast. Ion rocket motors get their speed by being constantly accelerated 24 hours a day for many months, and two versions of this technology are given for a low-power and high-power ion engine. Finally, solar sails can reach speeds of nearly that of the solar wind, and engineers are hopeful that this technology will be tested in space very soon. As you can see, we are currently stuck in the mode of travel where it takes nearly ten years to get to Pluto. Perhaps in another hundred years, this travel time will be reduced to a year or less.

344   Will humans ever travel to the stars?

After a lifetime of reading science fiction, I am very uneasy about answering this question based on how I feel today. It is with a heavy heart that I have to say, "Not anytime soon, or at least not within the next 200 years." There are many things that dictate this, the least of which is the technological issue of whether we can imagine ways to do it or not. These technical issues are very challenging. Even under the best of conditions, using our state-of-the-art ion engines, the trip to the nearest star with known planets would take hundreds of years. During this time, all systems would have to operate flawlessly, or be reparable. The travelers would have to recycle their air and grow their own food for a century. There would have to be some form of artificial gravity, and we really have no ideas what the long-term effects are of cosmic-ray exposure on humans. Our best opportunity is to let high-speed unmanned interstellar probes and remote sensing from the ground or near-Earth space give us the most accurate picture of how things are in the rest of the universe. We can then create a virtual-reality simulation of the whole shebang so that everyone on Earth can enjoy walking on the surface of Virtual Epsilon Eridani b, rather than just a handful of lucky astronauts. Our biggest problem is in reliability. There are no known examples of complex human technology that have survived without any repairs for more than 100 years. There are no service stations in space, so you will have to bring or manufacture all of your space parts during decades or centuries-long voyages. But the biggest problems facing us are not really technological at all.

All you have to do is look around you, and look at the trends in human behavior. We were on the Moon in the early 1970's and then abandoned the concept of a lunar colony when we had the best economic and political opportunity to get one going. We committed to building a space station, but no sooner did we spend the first $40 billion than politicians forced NASA to downsize from a seven-man to a three-man enterprise. We are now faced with the outright cancellation of the station because of a failure of political will, not technology. Space exploration costs money. It is the kind of investment that very few people nowadays seem to feel is justified. We had the makings of a lunar base and threw it away in 1972. Mars might be the exception, but politicians still can't be convinced there is a very good reason for going there, given a probable cost near $50 billion. Any human society that would underwrite an interstellar journey would look nothing like any society that now exists on Earth. That kind of society would be driven by a set of long-term imperatives that we can hardly imagine today. They would be impervious to weekly popularity polls or the quirkiness of a political system that elects a new administration every four years. Ancient Egyptians labored for decades, and by the tens of thousands, to build the pyramids that gave their divine pharaohs a royal send-off into the afterlife. Over a period of 100 years we built a trillion-dollar roadway and rail system. It is that kind of century-long focus that a society would need to make star travel happen.

345 If we never make it to the stars, what will humanity do instead?

It all depends on which way we want to extend our presence into the rest of the cosmos. If it is physically and economically prohibitive for any globe-girdling civilization to overcome the biological and technological issues surrounding interstellar travel, there is no loss of dignity associated with projecting our information-gathering technology into this grander arena. We are doing this already with unmanned interplanetary spacecraft and tele-operating on the surface of Mars. What we will probably do instead of interstellar travel is to send lightweight and fast probes to the stars where we are now detecting planets in orbit. That is why it's important to continue our current discovery process by remote-sensing distant planetary systems, and supporting ion-drive and solar-sail technology development. The catalog of planets we discover in the next ten years will be the catalog from which we select the destinations for future interstellar probes. They will use ion drives to travel at nearly half light speed and reach the nearest planetary systems in a century or less. As we widen this net of interstellar probes, we will have a steady stream of data flowing back every few decades. Of course, this has to assume that our civilization will persist long enough to continue listening for the signals from our own probes centuries after we have launched them.

346   If a marshmallow traveling at 99.99 percent the speed of light hit Earth, what would happen?

At this speed, special relativity says that the energy of this marshmallow equals about 100,000 terajoules. The most powerful American bomb, known as Castle/Bravo, was detonated on February 28, 1954, and released energy equivalent to an astounding 15 megatons of TNT or 84,000 terajoules. As with many uncontrolled natural phenomena, most of this energy will end up as heat, light, and sound, although there would also be tremendous ionization of the local atmosphere that would go along with this event. When you look at the energy per particle of the marshmallow, the energy is shared by 100 trillion trillion protons in the marshmallow for an average energy per particle of 10 billion volts per proton. This is more than enough energy to trigger some thermonuclear fusion and the production of gamma radiation. As for what it does when it actually impacts, it will cause a detonation equal in energy to a stony asteroid 200 yards across traveling at 20 miles/second and produce a crater nearly a mile across.

347   Are photons and gravitons really massless?

The latest measurement of the photon's mass was described by Alfred Goldhaber at the State University of New York at Stony Brook and Michael Nieto at the Niels Bohr Institute in Copenhagen in 1971. They used a variety of electrodynamic phenomena observed in nature to place stringent constraints on just how massive a photon could be without violating any known experimental result. If the photon contains as much as 2×10^{-47} grams of mass, departures from Coulomb's inverse square law for electrostatics would be detectable under laboratory conditions by a reasonably careful undergraduate student in a college physics laboratory. Even the shape of Earth's magnetic field would be measurably different if the photon were as fat as 10^{-48} grams. This mass limit is so absurdly small compared to other particle masses that a previously published estimate was found by Goldhaber and Nieto to be wrong by a factor of nearly one million using the same formulae and numbers. This mistake was never caught following its publication.

No one has ever seen a graviton, or any natural phenomena requiring such a particle. Its status is in even worse observational shape that that of the Higgs Boson. Goldhaber and Nieto also outlined the status of the search for a massive graviton ca. 1974. Although no quantum gravity theory was available, they decided that the next best thing to do was to establish whether any experiment could decide whether gravitons had to be massive or massless. To weigh the graviton, they used some of the largest collections of matter in the universe: clusters of galaxies, which are believed to be held together by the gravity of their constituent stars and galaxies. If gravity is carried by a massive particle, it cannot follow an inverse-square law behavior at all scales. At some critical distance it must begin to die off as the particles reach their maximum distances allowed by Heisenberg's uncertainty principle. In the clusters of galaxies Goldhaber and Nieto studied, the typical separations between galaxies were about 118,000 parsecs, with a maximum separation of 580,000 parsecs. Because of the apparent uniformity of the inverse-square behavior of gravity over these scales, this leads to an upper limit for the graviton mass of 2×10^{-62} grams. A limit, by the way, which is even more restrictive than the current best upper limit for the mass of the photon of 4×10^{-48} grams, also determined by Goldhaber and Nieto.

348   Has the decay of the proton been detected yet?

No. The best that can be done is to place lower limits to how fast it might happen. The current experimental limit in 2003 is longer than 6×10^{33} years based on a recent analysis of data from the Japanese Super-Kamiokande Neutrino Observatory.

349   What will space travel be like in the year 2264?

Contrary to what the science fiction TV series *Babylon 5* portrays, I reluctantly have to predict that, given what human society is preoccupied with these days, we will probably have working, stable colonies of a few dozen people on the Moon and Mars only by then. Tourism into low Earth orbit for $50,000 a trip will be thriving, just as some tourists can now visit Antarctica. The commercialization of the space within lunar orbit will be well under way. Mars will still be very exotic, but mainly a destination for a few dozen scientists per year looking for more fossils. I can imagine small spacecraft being sent into the Oort Cloud and perhaps beyond, but we are a long way from technology that lets us build machines capable of working for the hundreds of years needed to reach the nearby stars. I can imagine that the outer solar system will still be essentially unexplored by manned spacecraft because it is sooooo expensive. Plus, Europa aside, there is little need for humans to go out that far for adventure that can be supported by the rest of us through our taxes! I know this sounds pessimistic, but I think there will still be a lot of interesting things for us to do then. As for the actual technology, we will have fleets of single-stage-to-orbit space planes, solar sails, and ion engines. All of these have already been tested as prototypes. We just need to get ourselves motivated to be explorers and stop bean counting.

350 Why can't astronauts see stars outside Earth's atmosphere?

One big reason is that the portholes are not made of optical glass, so the feeble light from stars gets scattered and diffused in the glass. The same thing applies to the space-suit visors. Only cameras outside the spacecraft designed for low-light imaging would show stars. I discussed this with two astronauts: Harrison Schmidt (Apollo 17 geologist) and Ron Parise (shuttle mission specialist, STS-67). They both said that the stars are not easily visible through the shuttle portals or their space-suit visors. The blackness of space is unlike anything they have ever seen on Earth. They can see the Sun in the sky, and also some of the bright stars in the sky if they look away from the Sun. I imagine their view very much like what you would see on a mountaintop far away from city lights, with thousands of stars shining with a steady light, free of any atmospheric distortions. The intensity of familiar stars must be especially high because without an atmosphere, the light is more concentrated into point-like spots.

351 Will travel at light speed ever be possible?

No. Physicists, not science fiction authors, are the experts in boosting matter to nearly the speed of light, and it costs a pretty penny to do this even for puny little things like individual electrons and protons. A typical, commercial linear accelerator used for medical isotope production in hospitals boosts protons to modest energies of seven million volts. It costs just over $1 million but hardly gets protons to 1 percent the speed of light. Without our even thinking about relativity and what it says, the experimental fact is that to boost a single proton to a speed equal to 99.9999999999999 percent the speed of light would take an energy expenditure, somewhere in the machine, of 150,000 joules over the time it takes to accelerate the particle. That equals 40 kilowatt-hours, about the amount of electricity that a small house uses in one month. So, at this point you can do the math. Let's say a space shuttle has a mass of 200 tons. To boost this to a speed of even 99 percent the speed of light (which is fast enough to enjoy lots of time-dilation effects in interstellar travel) would require 2,000 billion billion joules of energy. This is equal to the power output of the Sun for a total of 60 hours. Even if you try to sneak up on this speed by accelerating for 50 years, you can't get something for nothing. At the end of the day, you will still have to cough up the 60 Sun-hours of energy. But a bigger problem is that, as you are cruising at near-light speed through the solar system and beyond, you are pelted by dust grains and micrometeors which now pass through your sorry butt like a bullet through cheese, kicking up lethal X rays as they hit your bulkhead—and your body—at near-light speed. One dust grain equals the destructive power of a small nuclear bomb.

352 How do you explain relativity theory to a 12-year-old?

I'm not sure you need to. A 12-year-old might realize that there are some interesting issues that have to do with space and time, but they can't separate scientific relationships between space, time, matter, and energy from the nonsense relationships they hear about every week. Nor should they be forced to. The scientific relationships are too subtle for children in that age group, and that's why educators don't make the attempt to describe relativity until the student has the necessary mathematical sophistication to understand the relationships in special relativity. This doesn't happen until college. None of relativity theory is intuitive even to adults. Your best bet is to say that nothing can travel faster than light and that objects that try to move really fast will be observed to run slowly, get smaller along their direction of motion, and gain mass. I think some things need to be left as a mystery to children. Until they learn the technical details in college, it is perfectly OK to let them enjoy the freedom of thinking imaginatively without trying to swamp them with facts. The language of spacetime and space warps they learn by listening to Jordie and Scotty on *Star Trek* is harmless, and it has inspired many of us to learn later on what is really going on by taking physics courses. Do not interfere with a child's ability to dream and explore ideas unfettered by factual relationships. This is a form of play that all professional astronomers practice as well.

353 According to relativity, how much younger are astronauts in Earth orbit after six months?

Let's assume that the astronauts travel at a speed near 20,000 miles per hour, which works out to about 0.00003 times the speed of light. According to special relativity, this produces a time dilation factor of 1.00000000045. If we multiply this by the six-month (15 million seconds) stay at the International Space Station traveling at the same speed, the astronauts would be about 15 million/1.00000000045 = 0.007 seconds younger than if they had stayed on Earth. Has this kind of thing ever been seen in real life? In 1971, Joe Hafele and Richard Keating at the U.S. Naval Observatory flew four atomic clocks on jet planes for 50 hours each. When they were brought back to the ground and compared, the westward-flying clocks had lost 273 nanoseconds and the eastward-flying clocks had lost 59 nanoseconds. These results were exactly what were predicted by relativity. So this kind of time travel is real and can make a big difference for such ultra-precise satellite systems as the Global Positioning System. This system would not work unless relativity were taken into consideration.

354   How are satellites prevented from crashing into each other?

Satellites stay on orbits that are very precise provided there are no external forces such as atmospheric drag to perturb them. The actual distances between most satellites are many hundreds of miles so there is no real danger of collision. However, there are now so many debris particles present in orbit that satellites, the space shuttle, and the space station run an increasing risk of being hit by orbiting debris from past space missions and exploded satellites. It is these uncontrolled small particles that cannot be avoided. U.S. Space Command can track objects the size of a walnut or a baseball in near-Earth orbit but cannot see ice droplets, flecks of paint, or nuts and bolts. These travel at thousands of miles an hour and are lethal. The Hubble Space Telescope has been hit, as has a space shuttle, and at least two satellites have been disabled after a debris collision.

355   What are normal satellite altitudes?

There is no normal altitude really. Satellites used for research can have orbits from a few hundred miles to tens of thousands of miles. Communication satellites are always found in the Clark Belt in geosynchronous Earth orbit (GEO) at an altitude of 35,786 miles. It all depends on what the satellite's purpose is. The Global Positioning System satellites orbit at 20,148 miles at mid-Earth orbit (MEO). Many newer communication satellite networks such as the 66-satellite Iridium network orbit at 483 miles, called low Earth orbit (LEO).

356   What are those strange lights I keep seeing in the sky?

This question, or specific versions of it, is very common to ask, especially if you are a casual observer of the sky as most urban people are these days. Even folks living in rural locations may not spend as much time looking at the night sky as their forebears may have. There are three classes of things to look for: fast movers, slow movers, and no movers. Fast movers that travel across the sky at the tempo of meteors or as fast as the second hand on your watch are, in all cases I have witnessed, simply meteors and satellites—or variations of these phenomena. It doesn't matter how bright they are or the color changes they go through. Meteors and fireballs are spectacular and have properties that vary almost on an individual basis. They can be colorful, sometimes audible, have bright tails or smoke puffs, or, for line-of-sighters, look like suddenly brightening stars that fade out in a few seconds. Sun glints off of satellites are also spectacular and especially common since 1996, when the Iridium network was put into place. Slow movers can be and usually are satellites, or distant jet planes' running lights, or, if near the horizon, mirages. Mirages can be seen at night if the atmospheric layer is lit at the right angle by distant car or city lights. As for no movers, if they are seen in a pair of binoculars or a telescope, they are probably geosynchronous satellites. They can be stars, but this is easily checked by looking an hour later and noticing that the light has actually moved. Near the horizon, the bright stars nearly always show frequent color changes due to turbulence and refraction of starlight in the atmosphere. Colored lights in the sky, if not star-like, can be caused by auroras if you look northward, or can be ice crystals scattering moonlight if you see a halo around the moon. There are noctilucent clouds that can have odd shapes in the early evening skies. And of course, there are many kinds of aircraft, civilian and military, which have complicated lights in various patterns, flying at many different altitudes. So, if you do see something, don't get too worked up about it, but look at it as an opportunity to learn something new about your nighttime world—a world that we safety-conscious, exhausted, urban dwellers rarely have time for anymore.

357 How can I locate an astronomer's E-mail address on the Internet?

I'm not allowed to say, but if you know the astronomer's name and you use the Google search engine, you might try doing a keyword search for his or her name and E-mail. I have found this to be 100 percent effective at least for reaching those astronomers who include their E-mail addresses in the personal biographical pages or the pages of their professional institutions.

358 Do astronauts need magnetic fields to survive in space?

No. There is no medical evidence I have ever encountered that says that the well-being of astronauts is improved by putting them in a large magnetic field. In fact, the instruments, experiments, and spacecraft systems would not tolerate having an unshielded magnetic field near them with a power strong enough to be therapeutic.

359   Why does the sky look black from space?

Because the camera is set to photograph a bright object like the Earth or Moon, and so it has to use a short exposure and fast f/stop. This is exactly the opposite of what you need to see stars in the sky, so the sky looks black. Figure 25 was taken on August 28, 1993, by the Galileo Mission as it flew by the asteroid Ida. Note the black sky surrounding the asteroid. You will also see this black-sky effect in Hubble Space Telescope images of Mars or even in photographs by ground-based telescopes. Go into an illuminated parking lot and look at the sky on a clear night. You will see the same effect as your eye is flooded by bright lights and "stops down." If you photograph a car illuminated by a streetlight, and then photograph the starry sky with the same camera settings, your photo will show few or no stars.

360   What are the Photon Belt, Nibiru, and Wormwood?

These are not astronomical objects but instead ingredients in a complex fantasy world in which some adults live. A simple search on the Internet will turn up thousands of references to these terms. What is a bit creepy is that these stories have caused their fair share of anxiety and even death among otherwise

FIGURE 25 Ida and its moonlet Dactyl imaged by the NASA Galileo spacecraft en route to Jupiter. The black-sky effect with no stars is common in all planetary photographs. (CREDIT: NASA/JPL-JHU)

sensible people. In these typically nonsensical discussions, it is claimed without any proof that we orbit the Pleiades every 24,000 years and pass through a region of intense radiation called the Photon Belt. It is claimed that this belt, also called the Golden Nebula by extraterrestrial aliens, causes spectacular changes in human spirituality every 2,000 years or so. Meanwhile, Wormwood was the name of a star that fell to Earth, as cited in the New Testament Revelations 8:10–11. In 1997, the Heaven's Gate cult in Rancho Santa Fe, California, believed that Comet Hale-Bopp was this prophesied star, and to reach the aliens inhabiting it, 39 members of this group committed suicide so that their spirits could be teleported to meet the aliens. Wormwood is also referred to by some believers as Nibiru or Planet X, the former name being taken from a mention in ancient Babylonian mythology. Some people even claim that Nibiru is the "fourth-dimensional flagship of the Milky Way Galactic Federation." Others claim that it is a body headed toward Earth with a powerful magnetic field that will destroy humanity in June 2003. These ideas only go to prove that human adults can talk themselves into some mighty strange beliefs, and force others to go along with them, often to their peril.

361 What is Ockham's razor?

First proposed by the medieval monk William of Ockham (1285–1347), this idea says that, given two explanations for something, Nature always seems to prefer the simpler of the two. What he actually said was, "Entities are not to be multiplied beyond necessity *(non sunt multiplicanda entia praeter necessitatem)*." I think it was made famous by the actress Jodie Foster playing Dr. Arroway in the movie *Contact*. What is amazing about this idea is that it isn't something we came up with and then applied to studying nature like some kind of philosophical corset. At many different levels, nature consistently shows us that it is very lazy. Water takes the simplest and most direct way down a slope. Electricity always takes the path of least resistance. Even the motion of planets and stars in gravitational fields always end up taking the straightest line they can between two points in space. The most successful theories we have are also the ones for which the number of extraneous assumptions are at the barest minimum. Ockham's razor is simply a restatement of nature's underlying simplicity, economy—and laziness.

362 What are some astronomical events that have affected history?

The kinds of things that can get humans really worked up are planetary conjunctions, solar eclipses, auroras, comets, meteor showers, and meteor falls.

Also, a few supernovas and novas are thrown in too since they light up the sky for days and weeks at a time. Not only have sightings of Halley's Comet been powerful portents of gloom and doom but certain planetary conjunctions have come at key times in human history. Table 12 provides a timeline of the many events that have had some historical impact since the earliest recorded sightings of solar eclipses. Humans have been terrified by spectacular meteor showers, and in some rare instances, have been injured or killed by being hit by meteorites. There are a number of total solar eclipses that were mentioned in ancient texts such as the one mentioned by Xerxes. The deaths of Herod, Augustus Caesar, and possibly Jesus Christ happened during lunar eclipses. When you tally up how many of these astronomical events have been recorded by humans and affected them in one way or another, the list is a very long one.

363 Will "tourist" space travel ever become practical or possible?

Probably not in your lifetime because of the way NASA gets funded by Congress. Now that we have experienced 14 astronaut deaths in two shuttle accidents, I think the public and NASA finally realize what scientists and engineers have been saying all along—that space travel can be very risky. The first space tourist, and probably the last for the foreseeable future, has already made the trip—Dennis Tito, a California millionaire, bought a $20 million ticket from the Russians for a six-day visit on the International Space Station in 2001. All this shows is that space tourism, if it ever succeeds, is once again another perc for the extremely wealthy—not for you and me. Aside from the safety issue for a fleet of shuttles now 20 years old, the biggest factor is the cost of launching a pound of payload (including rocket) into orbit. Right now, that number is near $10,000 per pound. NASA is committed to reducing this number to $1,000 in the next decade and to a few hundred dollars per pound by about 2020 using new-generation vehicle designs. Surveys, evidently, say that people are willing to pay up to $50,000 for a trip into space, and this works out to about $400/pound. The bottom line seems to be that in 40 years, launch vehicle costs could conceivably be so low that tourism will be ready to go and will be commercially lucrative. But visiting space will never be safer than taking a 9,900-mile flight from New York to Sydney, Australia.

364 What is the inside of an electron like?

The electron is one of the oldest-known fundamental particles, and a lot has been learned about it in the last 100 years. We know its mass (9.1095 ×

10^{-28} grams), its charge (4.8032×10^{-10} ESU), and its spin (1/2 units). From relativity experiments and the way its electromagnetic field behaves, we also know that it gains mass as it is accelerated in such a way that it cannot be a small sphere of matter with electrical charge on its surface. It is instead a truly point-like object whose "size" is smaller than 10^{-20} inches. Also, its relativistic mass increase is a result of its electromagnetic field energy, not of its rest-mass increasing. Rest-mass is a measure of a body's inertia when it is not moving in the reference frame of the experiment, and so it is the minimum possible mass that a body can have. Modern electroweak theory, which unifies the electromagnetic and weak-nuclear forces, says that electrons actually have no mass at all at energies of 100 GeV or higher. At these energies, the vacuum itself takes on a different energy state, and this causes the symmetry between electromagnetism and weak forces to break down. Electrons get their rest mass from the interaction of the massless electron field with the Higgs field in the vacuum. String theory says that electrons have no internal structure until you get to Planck energies of 10^{19} GeV, where their string-like characteristics become obvious. So, an electron is basically a knot of energy in the electromagnetic field, and the field itself is produced by the emission and absorption of virtual particles from within the electron itself.

365 What are some good web sites to visit if you want the latest astronomy news?

Since 1998, when I wrote *The Astronomy Café,* there has been a steady maturation of the Internet into an incredibly rich resource for astronomy and science reporting. You can start with the magazine news pages for *Sky and Telescope, Astronomy, Scientific American,* and *Physics Today,* and then pick up CNN's space science area at *http://www.cnn.com/TECH/space/.* The *New York Times* has a good general science section at *http://www.nytimes.com/pages/science/index.html.* Also, NASA has various science news pages featuring NASA discoveries at their main Headquarters page *http://www.nasa.gov/events/highlights/index.html,* or you can visit the Goddard Space Flight Center news page at *http://www.gsfc.nasa.gov/* or Science@NASA at *http://science.nasa.gov/.* For information about space technology and missions, space.com at *http://www.space.com* is a great resource. A good place to visit for an index of news features stressing planetary exploration is the JPL news notes archive at *http://www.jpl.nasa.gov/news/news_notes.cfm.* Another resource at Spaceweather.Com *(http://www.spaceweather.com/)* gives you the up-to-date story on the latest solar storms, sunspot counts, and near-earth asteroids. The Hubble Space Telescope news page at *http://hubblesite.org/newscenter/* will give you a roughly monthly update on the latest and greatest discoveries from that famous instrument, with dazzling images and wallpaper, too! Personally, I prefer visiting my library every week. I usually grab the latest copies of *New Scientist, Science News, Science,* and *Nature* and get up to date on the latest discoveries and issues. Every month I look at *Scientific American, Sky and Telescope,* and *Astronomy* magazines to get a larger perspective on major news stories. I also look forward to the weekly science sections of my local newspaper, the *Washington Post,* to follow the news stories that are considered to be of interest to the general public.

Glossary of Annoying Terms

accretion disk A flattened, rotating gas cloud usually found around a collapsed object such as a black hole, a neutron star, or a white dwarf. The gas has been "accreted" from some outside source such as a nearby companion star.

arcsecond There are 360 degrees in a full circle. If you take one of these degrees and divide it by 60 you get intervals of 60 arc minutes. If you divide each of these arc minutes by 60 you get arcseconds. There are 3,600 arc seconds per degree. An arcsecond is about what a toothpick's width looks like at 400 yards! It is also about how fat the image of a star looks from the surface of the Earth thanks to the turbulence in the atmosphere.

astronomical unit (AU) The distance between Earth and the Sun defined as 149,600,000 kilometers. It is a standard of measure for distances in the solar system. Also, it forms the basis for determining astronomical distances. At a distance of 3.26 light-years, one astronomical unit corresponds to an angular separation of one arcsecond.

barycenter The point in space around which two or more orbiting bodies in a multiple-body system orbit. The Moon does not really orbit around the center of Earth. Instead, both the Earth and the Moon orbit around a geometric point located a few thousand miles from the center of Earth. As seen from the Sun, the Earth-Moon system looks like a concentrated point of mass at the barycenter position, and it is that point that makes an elliptical orbit around the Sun every year.

brane According to some physicists, the space in our universe is a part of a three-dimensional object called a brane, which is in turn a part of a larger ten-dimensional space time in which the gravitational field operates. All other forms of matter and energy are restricted to our "3-brane," but there may be other 3-branes existing within the ten-dimensional spacetime that to their inhabitants would also look like complete universes, even infinite ones. Some physicists call these other 3-branes "braneworlds." Because the additional seven dimensions in space-time are very small, even sub-atomic, our 3-brane could be millimeters away from other 3-branes, and we would never know about this. All of our forces and forms of radiation (e.g. light) can only travel within our brane and not beyond.

conjunction   When a planet passes at its closest point in the sky to another planet or the Sun.

constant—cosmological   A factor added to some cosmological models to account for a possibly repulsive force exerted on matter by the physical vacuum. Also know as "Einstein's biggest blunder." As it turns out, it is a real effect now called dark energy.

constant—Hubble   The observable rate of increase of a galaxy's apparent velocity relative to Earth with distance, expressed in units of kilometers/sec per megaparsec. Its unit is actually 1/time, and cosmologists often refer to this as the Hubble Time. If the universe has been expanding at this rate since the big bang, this would be the age of the universe today, but in fact we know this is not the case because the expansion has been slowing down. Its current value is 71 km/sec/mpc.

Doppler shift   The shift observed in the sound or light signals from a body as it moves closer or farther away from you.

ecliptic plane   The band in the sky along which the planets and the Sun appear to move as viewed from Earth; a reflection of the fact that the orbits of the planets are confined to a very narrow range. Also the name given to the plane of the orbits of the planets.

event horizon   Every black hole is surrounded by a mathematical surface that separates the outside universe from the spacetime within the black hole. This surface is called an event horizon because events happening inside this surface can never be seen by outside observers no matter how much time passes. In a simple way, it is often said that an object just inside the event horizon will have to travel faster than the speed of light to escape the black hole. For black holes that do not rotate, the radius of the horizon surface—a perfect sphere—is 2.7 kilometers (1.7 miles) for every multiple of the Sun's mass that is inside the black hole. A 10 solar-mass black hole would have a horizon radius of 17 miles.

exoplanet   The name given to all of the new planets discovered to be orbiting other stars. Among these are the so-called epistellar jovians, which have about the mass of Jupiter and orbit their stars well within the size of the orbit of Mercury.

field   The specification of the intensity or other magnitude of a system at each point in space and time. On a conventional weather map, plotting the local temperature across the United States results in a temperature field. Plotting the arrows showing local wind speed gives you a velocity field. Fields are a very important concept in physics because at the most elementary level, all things including matter, forces, and energy have to do with a field, or can be described in terms of a field acting through space.

Heisenberg's Uncertainty Principle A principle in modern quantum mechanics that states you may not simultaneously make an infinitely accurate measurement of quantities such as the speed of a particle and its location. Perfect knowledge of one means complete lack of knowledge of the other. The consequence of this principle is that elementary objects such as electrons and subatomic particles obey laws of nature that are different than the ones we are familiar with. One of the most bizarre consequences is that empty space, a vacuum, is not really empty even if all the atoms are removed.

horizon—cosmological The limiting surface in three-dimensional space surrounding every observer defined by the distance light could have traveled toward the observer had it been emitted exactly at the instant of the big bang from a distant point in space. There is plenty of universe outside your local horizon, but you will not see the light signals from this distant matter until more time passes. This horizon expands at the rate of one light-year per year.

hyperspace The hypothetical collection of space dimensions beyond the three that we know in the physical world. This term is more often used by science fiction writers than by physicists, who prefer more specific terminology that actually specifies the number of dimensions involved.

mass A well-known but intrinsically mysterious property of matter that accounts for its resistance to being accelerated. In subatomic physics, some particles have zero rest mass, such as photons; others have a minuscule amount of mass, such as the neutrino, or enormous amounts, such as the W-bosons. Most of the measurable mass of a proton is an artifact of the energy of the gluon fields holding the three constituent quarks together, each of which has a rest mass of only a few percent of the total proton mass. Mass may be a property of the interaction of matter with a new field in nature called the Higgs field. Mass, by the way, is not the same as weight. Weight is the force produced by a specific amount of mass under acceleration.

matter—baryonic Matter that is in the form of "baryons," such as protons and neutrons.

matter—dark Matter that may exist in very large quantities in the haloes of galaxies, or in intergalactic space, which contributes to the gravitational mass of a system but not to the amount of light present in the system. There are two classes: baryonic and non-baryonic. The former includes dark matter consisting of black holes, faint stars, and other objects that produce little or no light detected from Earth. This kind of dark matter (about 4 percent of the universe), still consists of protons and neutrons, so it is called baryonic. The second category, non-baryonic dark matter, can include neutrinos that carry mass, or all other particles that produce no light but do not consist of protons and neutrons. The only known significant form of non-baryonic dark matter is neutrinos.

matter—missing   An older term used by astronomers to describe dark matter, both baryonic and non-baryonic, prior to the 1980's.

meteor   A body entering the atmosphere that produces a bright trail.

meteorite   A recovered body from the ground. This does not include rubble from man-made objects.

nebula   A cloud of gas in space that may be illuminated by a star, or be seen as a dark spot because no stars are nearby.

neutrino   A very light, possibly massless particle that only interacts with matter via the so-called weak nuclear force. Neutrinos can pass through many light-years of lead and only suffer a 50-50 chance of being absorbed.

neutron star   The remains of a massive star that has become a supernova. It represents the compressed core of the star and has about the mass of the Sun, but in a region only about 20 miles in diameter.

nova   The detonation of the surface of a collapsed object such as a white dwarf or neutron star due to mass flow from a companion star. The ignition of the fresh infalling fuel produces a million-fold increase in the electromagnetic output from the collapsed object.

nucleosynthesis   The process by which the heavy elements are created from the lighter elements, usually in the interior of massive stars or in the first few minutes after the big bang.

photon   The name given to the packet of energy of the electromagnetic field.

plasma   A gas composed of atoms that have had some of their electrons stripped. This "fourth state of matter" has nothing to do with the plasma in your blood or other fluids.

plenum   A general term that can be used to describe the "object" that underlies the physical world. It is not often used by physicists but can be thought of as synonymous with the gravitational field, spacetime, or any more primitive theoretical construct for these same physical things.

polarity   Opposing conditions such as "positive and negative," "north and south," "up and down," and "manic and depressive."

pulsar   A neutron star that is spinning and emitting pulses of electromagnetic energy at well-defined time intervals.

quantum   An irreducible packet of energy that has well-definable properties.

quantum field   A field consisting of individual particles whose interactions or properties lead to either specific forces or types of matter particles.

quantum jump   The change in a system from one stable state to another. Electrons can jump up or down in energy inside an atom if they are excited to do so by external influences. Not related to *Quantum Leap,* the TV series.

quark   Any one of the six fundamental particles that combine to make protons, neutrons, and other heavy nuclear particles that primarily interact via the strong nuclear force.

quasar   A class of distant galaxies that have a central energy source, probably a supermassive black hole, which is emitting tremendous amounts of electromagnetic energy.

radiation   A loose term used to describe the flow of energy in space from a source. It can be a stream of particles, but astronomers generally reserve the moniker "radiation" for electromagnetic energy or gravitational energy.

radioactive dating   Certain types of elements called isotopes literally fall apart after a set amount of time. Their nuclei fragment into simpler and more stable forms by emitting neutrons or protons, for example. Different isotopes decay to more stable elements at different rates called their half-lives. Scientists can use this measurable timing to "date" how old rock samples are. If you start with 100 atoms of an isotope, in a time equal to its specific half-life, this isotope will end up as 50 stable atoms of the element and 50 isotopic atoms. After a further elapse of a second half-life time, you will end up with 75 stable atoms and 25 isotopic atoms. For a rock sample where both the stable and isotopic forms have remained trapped since solidification, you can accurately calculate the age of the rock. Sometimes the same rock can be dated using several different and independent nuclear chronometers, and you end up with very consistent ages. Astronomers have even used this method to date a star. (See Question 165.)

red shift   The shifting of a sound or light signal to longer "red" wavelengths. In cosmology, it refers to the phenomenon in which the space between galaxies has expanded, and this expansion has stretched the wavelength of light emitted by distant galaxies so that they now are longer than at the time of emission.

singularity   A mathematical condition in which some physical quantity takes-on an infinite or undefined value. Specifically in general relativity, the state of infinite spacetime curvature at the core of a black hole.

solstice   The time during the planetary year when the Sun is at its greatest distance from the ecliptic plane as seen from the surface of the planet. For Earth, these dates are near June 21 and December 21.

space   The property of the physical world embodied in the separations between objects.

string theory   This is a mathematical description of matter, field, and space that was first proposed in the early 1980's by Michael Green (University of London), John Schwarz (Caltech), and Edward Witten (Princeton University). The basic idea is that all of the known particles and fields are actually composed of minute loops of string that exist in a 10 or 11-dimensional arena that is more basic than our familiar spacetime. The extra dimensions are presumably the size of the so-called Planck scale (10^{-33} centimeters) and contain information that discriminates one kind of string from another as to its physical properties (spin, mass, type, etc.). When combined with a principle called supersymmetry, one kind of particle can be mathematically converted into another, which means that particles and fields are completely unified. Also, one obtains a model for the gravitational field in terms of these "superstrings," which can also be unified with the other three forces of nature, so the theory is completely unified with gravity in a simple and beautiful way, mathematically. There were five distinct superstring theories known by 1995, and these actually are unified into one even larger theory called M-theory. An important verification of string theory will involve the discovery of supersymmetry in the next generation of high-energy physics experiments. Superstring theory offers astronomers and physicists an explanation for dark matter and dark energy, which are ingredients to the universe not explainable by the current standard model of high-energy particle physics.

supernova   The detonation of a massive star that typically produces a luminosity increase equal to several billion times that of the Sun for a several-month period.

tidal effects   The many effects upon a body due to the gravitational influences upon it by a nearby body. Usually expressed in terms of a deformation in shape.

tunneling   The quantum mechanical ability of a system to change from one state to another in a way that is forbidden by the laws that operate on macroscopic particles and systems. This ability is often expressed by the statement that an electron can escape from any box, or it can escape from any confining energy potential by "tunneling" though an energy barrier or potential. When radioactive atoms decay, they do so because some of the neutrons and protons in their nuclei can escape from the nuclear "well" by tunneling to the outside.

Universal Time   The local civil time measured at an Earth longitude of zero degrees located in Greenwich, England.

vacuum   A poorly understood physical condition in which no physical object appears to be present, but in which virtual particles abound and can produce some astounding phenomena despite the fact that they are not directly observable. According to physicists these come in three types "true," "false," and "Hoover."

Table 1 : Atmospheric Meteor Detonations

Date	Yield	Location
June 30, 1908	20 megatons	Siberia
August 3, 1963	1 megaton	Africa
September 26, 1962	20 kilotons	Mideast
September 27, 1962	30 kilotons	Mideast
March 31, 1965	20 kilotons	Revelstoke
February 1, 1994	40 kilotons	Marshall Islands
August 11, 1998	9 kilotons	Australia
January 18, 2000	5 kilotons	60N 225E
February 18, 2000	15 kilotons	0.86N 109.1E
April 23, 2000	8 kilotons	Pacific Ocean
August 25, 2000	3 kilotons	Mexico

NOTE: 15 kilotons of TNT equals a single "Hiroshima Bomb" event. For more information read the February 1994 *Astronomy* magazine article by Christopher Chyba. The objects range in size from 6-feet to 50-feet in diameter. The 1908 event was the Tunguska Impact. The 1963 was called the "Bay of Pigs Event."

Table 2: Upcoming Asteroid Close Approaches

Name	Distance	Close Approach
2002 CA26	612,000 miles	February 5, 2002
2002 CB26	298,000	February 8, 2002
2002 EM7	289,000	March 8, 2002
2002 GQ	262,000	March 31, 2002
2002 FD6	674,000	April 6, 2002
2002 NY40	327,000	August 18, 2002
2002 TY59	474,000	October 1, 2002
2002 TZ66	617,000	October 5, 2002
2002 XV90	73,000	Dec 11, 2002
2001 WN5	137,000	June 26, 2039
1998 OX4	186,000	January 22, 2148
2000 WO107	216,000	December 1, 2140
1999 AN10	242,000	August 7, 2027
35396	251,000	October 28, 2136
1999 RQ36	325,000	September 23, 2080
2001 GQ2	362,000	April 27, 2100
2002 CU11	372,000	August 31, 2080
2001 WN5	400,000	June 25, 2028
1999 RQ36	465,000	September 23, 2060
2002 NN4	483,000	June 7, 2130
1999 RQ36	511,000	September 25, 2158
2002 AW	520,000	October 6, 2103
Hathor (2340)	548,000	October 21, 2086
1999 VP11	567,000	October 22, 2086
35396	576,000	October 26, 2028
2000 TU28	604,000	April 14, 2121

Table 3: The Farthest Outposts of the Solar System

Name	Distance
1998 XY95	65.1
Pioneer 11	66.0
Voyager II	71.1
Pioneer 10	82.0
1996 TL66	84.2
1999 CF119	90.6
Voyager I	91.1
1999 CY118	91.6
1996 GQ21	93.6
1999 RU214	95.5
1999 TD10	97.6
1999 RZ215	100.0
2002 GB32	100.0
1999 CZ118	116.0
1999 RD215	118.0
2000 PJ30	119.0
2001 YH140	197.0
2002 VQ94	216.0
2001 FP185	226.0
2000 CR105	230.0
1996 PW	279.0
2000 OO67	542.0

NOTE: Distances for the Pioneer and Voyager spacecraft are estimated for 3/2003. All distances are given in Astronomical Units. Pluto's distance, by comparison, is 40 AU.

Table 4: The Largest Known Stars

Name	Mass	Size	
VV Cephi	——	3,700	(Orbit of Uranus)
Mu Cephi	——	1,500	(Orbit of Saturn)
Betelgeuse	——	900	(Orbit of Jupiter)
Antares	——	530	(Orbit of Mars)
Deneb	——	145	(Orbit of Venus)
Pistol Star	200	——	
Eta Carina	150	——	
LBV 1806-20	>120	——	
HD 93129A	120	——	
Cyg OB2/ 12	92	——	
Zeta Scorpii	60	——	
P Cygni	60	——	

NOTE: The masses and sizes are given in multiples of solar units.

Table 5: Estimates for the Mass and Star Content of the Milky Way

Date	Reference	Mass	Stars
1978	Shipman, *The Restless Universe*	200 billion (total)	——
1980	Lang, *Astrophysical Formulae*	150 billion (interior)	——
1991	Zeilik, *The Evolving Universe*	720 billion (total)	——
1994	Douglas Lin (UCSC)	600 billion (total)	——
1996	Sky and Telescope (6/1996)	490 billion (total?)	——
1997	Barbara Ryden (course: Ohio State)	93 billion (interior)	——
1997	*Ask the Astronomer* (AstronomyCafe)	120 billion (interior)	240–1,200 billion
1998	SEDS web site	750–1000 billion (total)	——
2000	Shawl, *Discovering Astronomy*	200 billion (interior)	200 billion
2000	*The GAIA Study Report*	2 trillion (to 300kpc)	——
2002	Gene Smith (UCLA tutorial)	——	400 billion
2002	Imamura (course: Univ. Oregon)	600 billion (total)	400 billion
2002	*Ask the Astronomer* (Cornell U.)	1 trillion (total)	——
2002	G. Bell (Harvey Mudd U.)	700 billion (total)	——

Table 6: Conversion of Red Shifts to Cosmological Time and Space Measure

Z	Distance	Light Travel Time	Time Since Big Bang	
1	10.8	7.7	5.93	billion
2	17.1	10.3	3.34	billion
3	21.1	11.5	2.19	billion
4	23.8	12.1	1.57	billion
5	25.9	12.5	1.19	billion
6	27.5	12.7	950	million
7	28.8	12.9	780	million
8	29.8	13.0	650	million
9	30.7	13.1	560	million
10	31.5	13.2	480	million
15	34.2	13.4	270	million
20	35.9	13.5	180	million
50	40.0	13.6	47	million
100	42.2	13.65	17	million
1000	45.6	13.7	430,000	

NOTE: Distance is given in units of billion light-years. Travel time is in billions of years, and time since the big bang is in years. Obtained from current "Standard" cosmological model based on Omega (dark energy) = 0.73, Omega (dark matter) = 0.23, H0 = 71 km/sec/mpc and universe age 13.7 billion years. See Ned Wright's *Cosmology Calculator* at http://www.astro.ucla.edu/~wright/Cosmo Calc.html

Table 7: The Most Distant Objects in the Universe

Object	Date	Type	Z	Distance	Age
HCM 6A	2002	Galaxy	6.56	27.6	0.8
TN J1338-1942	2002	Cluster	4.1	23.6	1.4
Lens Abell2218	2001	Galaxy	5.58	26.3	1.0
SDSS 1044-0125	2000	Quasar	5.82	26.6	0.9
SN1997ff	1997	Supernova	1.7	15.3	3.7
CL1358+62	1997	Galaxy	4.92	25.2	1.2
TN J0924-2201	1998	Galaxy	5.19	25.6	1.1
PC1247+3406	1991	Quasar	4.897	25.2	1.2

NOTE: Distance given in billions of light-years. The time since the big bang (age) is given in billions of years.

Table 8: The Cosmic Timeline

Time	Event and Comments
10^{-43} sec	Planck Era. String World. Birth of time and space.
10^{-36} sec	GUT Era. Strong and weak forces freeze-out.
10^{-34} sec	End of Inflationary Era.
10^{-23} sec	Hadron Era. Massive particle soup.
10^{-12} sec	Four forces. Temperature = 10,000 trillion K.
10^{-6} sec	Quark freeze-out. Temperature = 10 trillion K.
10^{-4} sec	Lepton Era.
0.01 sec	Nucleons appear as free particles.
1 second	Radiation Era. Universe transparent to neutrinos. Temperature = 1 billion K.
3 minutes	Nucleosynthesis era. Helium, deuterium, berillium, and lithium appear.
1 hour	Free neutrons vanish from the cosmos.
300,000 years	Radiation Era ends. Cosmic background radiation. Temperature = 3,000 K.
500,000 years	Cold Universe. Dark Ages begin. No stars yet.
300 million	Population III stars start to form. Heavy elements start to appear.
400 million	Hydrogen in cosmos is reionization by Pop III stars.
600 million	Objects similar in mass to Large Magellanic Cloud become common.
700 million	Formation of oldest stars in Milky Way such as 72 Herculis and HD 221830.
850 million	Dark Ages have ended.
900 million	Galaxy 53W091 already has stars 3.5 billion years old.
13 billion ago	Hydrogen is fully reionized by this time.
12.5 billion	Formation of Milky Way's disk population. Oldest dated star: CS 31082-001.
12 billion	The oldest pulsar PSR J2145-0750 is born.
11.5 billion	Most stars in elliptical galaxies and the bulges of spiral galaxies have formed.
8.8 billion	The universe begins period of accelerated expansion.
8.0 billion	A burst of star formation activity "Burst C" occurs near solar neighborhood.
7.0 billion	The star Beta Hydae forms.
6.8 billion	Formation of Alpha Centauri A.
6.0 billion	Formation of oldest detected white dwarf. PSR J2145-0750.
4.9 billion	Murchison Meteorite forms.
4.6 billion	Tagish Lake meteorite. "Primitive" organic chemicals used to build amino acids.
4.59 billion	Aluminum-26 isotopes pollute the solar nebula from an ancient star.
4.58 billion	Max age of solar system. HH 237 and QUE 94411 with iron chondrules.
4.57 billion	Weak-lined T Tauri. Jovain planet formation ceases. Terrestrials still forming.
4.56 billion	Active T-Tauri Phase Flares 100 times more luminous and frequent than now.
4.55 billion	Core of Earth formed within 50 million years of Solar System.
4.54 billion	Age of Earth based on the Canyon Diablo meteorite and zircons.
4.5 billion	Age of Martian meteorite ALH84001.
4.5 billion	Oldest Moon rocks. The lunar highlands are formed.
4.5 billion	Moon formation within 100 million years after the solar system's birth.
4.4 billion	Oldest rock. Zircon crystal. First crust. Sun spins once in 45 days.

Time	Event and Comments
4.35 billion	Mars likely has dense primordial atmosphere during formation.
4.3 billion	Earth has liquid water.
4.0 billion	ALH84001 ejected. Sun 25% dimmer. Hemoglobin appears.
3.8 billion	Hadean Era ends formation of the earth. Archaean Era begins.
3.7 billion	Heavy Bombardment Era ends. Martian atmosphere stops evolving.
3.6 billion	Formation of Kaapvaal Craton in Africa. Oldest surviving crust.
3.5 billion	Cyanobacteria appear and respire oxygen. Earth day is 15 hours long.
3.2 billion	Moon orbits Earth in 20-day month. 550 days = 1 year.
3.0 billion	Proterozoic Era. Ancient life-forms look remarkably similar to modern algae.
2.6 billion	The end of Archaean Era. Atmosphere methane, ammonia, stromatolites common.
2.5 billion	Proterozoic Era. First abundant living organisms, mostly bacteria and archaeans.
2.3 billion	First Ice Age. Continents near equator glaciated. Earliest known Snowball Earth.
2 billion	Thick extensive "red beds"; free oxygen present; nitrogen 96% of atmosphere.
2.02 billion	Vredfort crater in South Africa, 300 km in diameter, formed.
2 billion	Mars atmosphere thins. Liquid surface water begins to disappear.
2 billion	Eukaryotic cells assemble from distinct organisms. Aerobic respiration.
1.85 billion	Sudbury Crater in Canada, 250 km across, formed by asteroid.
1-2 billion	Cyanobacteria dominate. Eukaryotes went nowhere evolutionarily.
1.9 billion	Pre-Rodinia supercontinent "cratons" seen in Australia and North America.
1.6 billion	Animals, plants, and fungi diverge from a common genetic ancestor.
1.4 billion	Fragmentation of Paleoproterozoic supercontinent.
1.1 billion	Late Proterozoic Eon. Supercontinent of Rodinia assembled.
1 billion	Oldest fossil raindrops. Milky Way's young disk population of stars forms.
900 Mya.	Length of day = 18 hours so 1 year = 481 days.
800 Mya.	Genome Alu repeat element "house cleaning" by mammal ancestor.
750 Mya.	Supercontinent of Rodinia breaks apart. Ice covers Earth to Tropics.
700 Mya.	Land plants and fungi evolve. Praesepe star cluster in Cancer forms.
680 Mya.	Ancient solar activity cycles in progress. Weakening of the geomagnetic field.
660 Mya.	Hyades cluster forms.
650 Mya.	Late Proterozoic. A major ice age. Precambrian great extinction.
600 Mya.	Mars climate system collapses by this time.
575 Mya.	Snowball Earth. Phase may have occurred. Mass extinction.
555 Mya.	The White Sea of Russia's Ediacaran fauna appear.
550 Mya.	Break-up of an older supercontinent prior to Pangaea.
543 Mya.	At least four major extinctions occurred during the Cambrian. Earliest date.
530 Mya.	Earliest fish fossil found in Yunnan China. Myllokunmingra.
500 Mya.	Global resurfacing event occurs on Venus.
500 Mya.	Oldest footprints ever found of centipede-like creature.
440 Mya.	Ordovician extinction. 100 families of marine invertebrates perished.
425 Mya.	Tropical fossils found in Scandinavia and northern Appalachian region.
420 Mya.	O_2 and O_3 increase. Surface UV intensity tolerable for living systems.
408 Mya.	The Devonian mass extinction occurs.
350 Mya.	Extensive glaciation of Africa, South America, India, Antarctica, Australia.

Time	Event and Comments
440 Mya.	Late Ordovician glaciation.
400 Mya.	Oldest insect known; winged Rhyniella precursor. Praesepe star cluster born.
390 Mya.	Supercontinent Gondwana. A vast ocean covers the rest of the planet.
370 Mya.	Archeopteris is the first woody tree to begin to flourish.
330 Mya.	The star Sirius is born.
320 Mya.	Mammalian ancestors show up.
310 Mya.	Genetic divergence of birds and mammals begins.
306 Mya.	Pangaea being built. Long-lived Ice Age climate conditions.
300 Mya.	Carboniferous Era. Xi Leporis form. Near-complete resurfacing Venus.
255 Mya.	Pangaea surrounded by a single ocean, the Panthalassic. Desert conditions.
251 Mya.	Fullerenes suggest major asteroid impact triggered mass extinction.
250 Mya.	Flowers evolve.
248 Mya.	The Permian mass extinction greatest ever recorded.
230 Mya.	Mammals are relatively rare, small in size, and likely nocturnal.
208 Mya.	The first true mammals appear in Late Triassic, and persist to the present.
200 Mya.	Cartwheel Galaxy forms. Triassic-Jurassic Extinction. Magellanic Stream forms.
195 Mya.	The breakup of Pangaea is just beginning.
156 Mya.	Massive releases of methane from gas hydrate layers beneath ocean floor.
140 Mya.	Angiosperms explode and become dominant plant forms.
100 Mya.	The Pleiades star cluster has begun to form.
92 Mya.	Oldest ant discovered in amber.
80 Mya.	Sirius-B evolves into a white dwarf.
64.98 Mya.	The Chicxulub Crater in Mexico, 170 km, formed by asteroid 10–15 km.
65 Mya.	Deccan Trap lava eruption covers India, floods earth's surface with CO_2.
65 Mya.	Cretaceous Extinction: Eighty-five percent of all species disappear.
60 Mya.	Primates begin to develop from insectivores in a forested environments.
57 Mya.	Eocene Era: Trees grow in Arctic and Antarctic.
58 Mya.	Earth's climate warms considerably. Intense tectonic activity.
55 Mya.	A tremendous release of methane gas frozen beneath the sea floor heats Earth.
55.5 Mya.	Many species appear including ancestors of hoofed animals.
54 Mya.	Early Eocene era is about 5.5 C warmer than the present.
50 Mya.	Dramatic decrease in genetic repeats in the human genome begins.
35 Mya.	The Popigai Crater in Russia, 100 km across, is formed by asteroid impact.
35.5 Mya.	Chesapeake Bay Crater, Virginia, USA, 85 km.
36 Mya.	First of three major cooling steps. North American air drops 12 degrees Celsius.
30 Mya.	Human retrovirus HERV-K infects Old World; human ancestors.
23 Mya.	Onset of major ice cover on Antarctica as orbit eccentricity drops to zero.
15 Mya.	The Martian meteorite ALH84001 kicked off Mars and ejected into space.
12 Mya.	The Sco-Cen OB supernova begins to clear out Local Bubble.
10 Mya.	Million-degree gas in the Local Bubble produced by three supernovae.
9 Mya.	Ancient ape Orepithecus babolii has an opposable thumb.
6.5 Mya.	Divergence of Pongid "ape-chimp" and hominid lines based on DNA studies.
5.8 Mya.	Mediterranean basin becomes an arid desert through continental drift.

Time	Event and Comments
5.7 Mya.	*Ardipithecus ramidus* appears in Ethiopia's Middle Awash region.
5 Mya.	Sco-Cen OB supernova deposits Fe60 and Ni60 in ocean sedimentary deposits.
5 Mya.	Atlantic Ocean bursts though the Gibraltar Straite like thousands of Niagaras.
5 Mya.	Left-hemisphere dominance appear in chimps, gorillas.
5 Mya.	Earth warms. Trees grow in Greenland and Canada as far north as 82 degrees.
3.5 Mya.	Bipedialism begins to emerge. *Australopithecus africanus.*
3 Mya.	First stone tools found in Omo area of Ethiopia. Eta Carina forms.
2.1 Mya.	Huckleberry Ridge caldera explodes in Yellowstone Park.
2 Mya.	Human DNA acquires freeloading parasites seen in Alu sequences.
2 Mya.	African "bottleneck"–triggered genetic changes start evolution of our species.
2 Mya.	Supernova in Sco-Cen OB damages ozone layer and causes extinctions?
1.8 Mya.	First mammoths enter North America from Eurasia across Bering Strait.
1 Mya.	Center of 3C295 wracked by an awesome explosion.
980 Kya.	End of the Jaramillo Normal Polarity Chron.
930 Kya.	Beginning of the Matuyama Reversed Magnetic Polarity Chron.
840 Kya.	Humans migrate to Flores, implying seafaring technology.
800 Kya.	Prolonged childhood detected in teeth of hominid fossils in Spain.
760 Kya.	Long Valley Caldera blankets entire western United States in ash.
750 Kya.	Earth's magnetic field finishes its transition to the current polarity configuration.
500 Kya.	Fire first used in China, Hungary, and France.
400 Kya.	Language emerges around this time. Humans use paint for decoration.
340 Kya.	Geminga Supernova explodes about 300 ly from Earth.
300 Kya.	Orion Nebula Trapesium stars form and light up the Orion Nebula.
240 Kya.	Illinoian Glaciation buries Canada and the northern United States for millennia.
200 Kya.	Out of Africa bottleneck for *Homo sapiens* lasts perhaps 200 years.
125 Kya.	Eemian Interglacial era. Global climate warmer. Sea level 5 or 6 meters higher.
120 Kya.	Human adolescence may have emerged by this time.
100 Kya.	Mars polar deposits approximately this age since no large craters present.
100 Kya.	Neanderthals use identical tool "kits."
100 Kya.	*Helicobacter pylori* takes up residence in human stomach by at least this time.
100 Kya.	Modern humans contemporaries of Neanderthals appear.
89 Kya.	Genetic study of Oceanic populations shows a common ancestor left Africa.
77 Kya.	Oldest cave art. South Africa. Blombos Cave.
74 Kya.	Toba, in Sumatra, erupts explosively. Major climatic cooling event ensues.
70 Kya.	A human population crash occurs worldwide.
63 Kya.	Hyoid bone in Neanderthal skull. Language and complex vocalizations possible.
60 Kya.	*H. Sapiens* found in Australia. "Speciation event" permits modern behavior.
53 Kya.	Northern Europeans descended from 50 individuals who survived last ice age.
49 Kya.	Barringer Crater is formed by an asteroid impact in Arizona.
46 Kya.	Australian large mammal and bird extinctions.
45 Kya.	Cro-Magnon man: stone-working methods; Neanderthals finally yield Levant.
40 Kya.	"Population bottleneck"—number of humans falls to a low level.
40 Kya.	*H. Sapiens* enter Europe. *H. neanderthalensis* still present. First Venus figures.
40 Kya.	The 7R allele of gene ADHD first enters the genome.

Time	Event and Comments
33 Kya.	Isotopes in Antarctic ice record associated with Local Fluff cloud complex.
32 Kya.	Engraved bone flute from France; new "sensibility" appearing in early humans.
30 Kya.	Neanderthal vanishes abruptly in Iberian Peninsula.
30 Kya.	Amerind people arrive in North America.
30 Kya.	Paintings in Chauvet Cave.
29 Kya.	Youngest Neanderthal is found on slopes of Caucasus Mountains in Russia.
25 Kya.	Gravettian migration of people into Europe from the Tigris-Euphrates region.
24 Kya.	Climate worsens. Gravitteans move to Balkans.
20 Kya.	Northern continents under ice 3 km thick. Sea level 130 meters below current.
20 Kya.	Ancient lunar phase calendar. Scratched lines and holes in sticks and bones.
20 Kya.	San Francisco Bay is a valley. Sea level 150 meters lower.
20 Kya.	Cygnus Loop Supernova.
18 Kya.	Wisconsinan Ice Sheet spreads to Pennsylvania, Ohio, Indiana, Illinois.
18 Kya.	Clovis People may have lived in the Cactus Hill, Virginia, area.
16.5 Kya.	First star maps in Lascaux. Pleiades, Summer Triangle, Northern Crown.
14 Kya.	Bering Land Bridge is largely flooded. Migrations cease.
14.7 Kya.	The Monte Verde campsite in Chile inhabited. About 30 people live there.
13.5 Kya.	Oldest human skeleton from Brazil resembles African and aboriginal stock.
10000 B.C.	Most recent Magdalenian cave paintings. Earth warmer. Ice sheets melt.
9760 B.C.	Malaria passed to humans. Malaria resistance gene G6PD appears.
9600 B.C.	Younger Dryas. Decline in plant and animal resources for humans.
9000 B.C.	The oldest known camps have been found along the ice-free corridor in Canada.
9000 B.C.	Rise in temperature in just 40 years.
9000 B.C.	Spirit Cave first occupied in northeast Thailand.
9000 B.C.	ALH84001 lands on Antarctica after being in space for the last 15 million years.
8300 B.C.	End of Pleistocene based on varv chronology.
8000 B.C.	Vela Supernova erupts. Gum nebula is formed.
8000 B.C.	70 North American species disappear, three-quarters of them large mammals.
8000 B.C.	Current Holocene Interglacial Era. First plants and animals domesticated.
8000 B.C.	Jericho has 2,000 inhabitants.
8000 B.C.	Koelbjerg Woman, oldest bog body known.
8000 B.C.	Human population size estimates at a few million.
7300 B.C.	Kennewick Man, similar to Europeans and Middle Easterners in N. America.
7000 B.C.	Cultivation of rice, taro, and yams in Asia. Atacama mummies.
6000 B.C.	Rock art in the El-Hosh area of upper Egypt.
5650 B.C.	Black Sea Flood? Bosphorous catastrophe supported by flow calculations.
5500 B.C.	Maritime archaic burial mound, L'Anse Amour, Labrador.
5500 B.C.	Agriculture, animal husbandry, and Indo-European languages emerge.
4500 B.C.	Tomb building. Sumeria is the first to begin this process of construction.
4236 B.C.	Earliest Egyptians devise a 365-day calendar that begins this year.
3800 B.C.	Sighting of the Campo del Cielo meteor impact in Argentina.
3500 B.C.	Sahara desert created in sudden climatic "flip," devastating ancient civilizations.

Time	Event and Comments
3500 B.C.	First writing, Sumerian.
3500 B.C.	Ring Nebula in Lyra probably forms at this time.
3400 B.C.	Invention of Wheel—pictographic sign, and turned pottery is common.
3300 B.C.	Recovery of Ice Man from an alpine glacer.
3100 B.C.	Egyptians begin use of hieroglyphic writing.
3100 B.C.	Stonehenge I is built.
3000 B.C.	Bronze Age begins.
2900 B.C.	First Dynasty of Egypt founded.
2767 B.C.	Earth's oldest living inhabitant "Methuselah" Bristlecone Pine in Inyo Mountains.
2650 B.C.	Li Shu begins recording astronomical observations in China.
2500 B.C.	First written story of the Flood by the Sumerians.
2357 B.C.	Chinese mention the Pleiades star cluster.
1900 B.C.	First Akkadian accounts of a Deluge. Hero is Atrakhasis.
1628 B.C.	Minoan civilization falls after Thera volcano on Santorini explodes.
1660 B.C.	Santorini explosion. Implications for Exodus biblical history.
1615 B.C.	The fall of Jericho based on radiocarbon dating.
1420 B.C.	Meteorite fall. Joshua 10:11. Lethal stones were cast down from heaven by God.
1375 B.C.	Earliest solar eclipse recorded in Babylon. Ugarit on May 3, 1375 B.C.
1335 B.C.	Exodus? Egyptian records—no evidence. Short reign of Ramses I (c. 1335–1333).
1348 B.C.	Ramses II, builder of Pithom and Raamses.
1277 B.C.	Exodus? Assyriologists and Egyptologists favor this date.
1200 B.C.	Babylonians write *Gilgamesh*, *Enuma Elish*, and *Atrakhasis*.
1200 B.C.	Oldest date for early Olmec civilization near San Lorenzo in Veracruz, Mexico.
1059 B.C.	Possible first sighting of Halleys Comet recorded. Halley's Comet has.
747 B.C.	Babylonian cuneiform tablets record astronomical tables on a regular basis.
300 B.C.	Mayan pyramid temple of Uaxactun built.
1 A.D.	Estimated population is at 150 million people.
33	A proposed year for the crucifixion of Christ from eclipse.
616	First known death-by-meteorite occurs in China.
900	Collapse of Mayan civilization due to drought.
900	The Medieval Warm Period. Viking civilization begins its expansion.
1006	Supernova seen in China, Japan, Europe. Identified with a radio SNR.
1054	Crab Supernova seen in China and Japan.
1276	Possible end of Anasazi Civilization. Severe drought lasting 20 years.
1290	Wolf Minimum of solar activity begins based on auroral sightings.
1420	Sporer Sunspot Minimum begins based on C14 abundances.
1492	The oldest meteorite fall witnessed occurs in November 1492 in Ensisheim.
1600	CO_2 concentration at the 250 micro-liter/liter level for past recorded history.
1664	Jupiter's great red spot first seen/acknowledged by Robert Hook.
1700	Estimated population is at 600 million people.
1730	Anders Juhlin, great-great-great-great-great grandfather to Sten Odenwald, born.
1815	Eruption of Tambora is the largest explosive eruption in recent historic time.

Time	Event and Comments
1843	Eta Carina outburst.
1848	Auroral currents detected in telegraph lines. Great Aurora of November 17.
1900	Estimated population is at 1.6 billion people.
1920	Dow Jones Industrial Average is at 80 points.
1950	Rates of extinction for birds and mammals has increased significantly.
1950	HIV-1 crosses into humans, possibly from chimpanzees.
1952	Author Sten Odenwald is born.
1957	Soviets launch Sputnik.
1960	Atmospheric CO_2 at 315 parts per million.
1961	First human to enter space.
1969	Humans set foot on the Moon for the first time.
1980	Atmospheric CO_2 is at 335 parts per million and is continuing a linear rise.
1980	Global temperatures +0.2–0.3 C warmer than the pre-1930 era.
1985	Maser source K3-35 shows the first signs of becoming a planetary nebula.
1987	Supernova seen in Large Magellanic Cloud. Neutrino pulse arrives as well.
1987	Dow Jones Industrial Average is at 2,000 points.
1990	Atmospheric CO_2 is at 352 parts per million.
1990	Increasing reports of a disappearance of amphibians in North America.
1990	Global temperatures now running 0.5 C higher than pre-1930 climate.
1994	Satellites damaged by solar storm. ANIK E-1 and E-2.
1995	Discovery of first planets orbiting a star—a pulsar.
1998	Warmest year of the millennium recorded.
1998	August 27. Gamma-ray burst lights up atmosphere. SGR 1900+14.
1999	Dow Jones Industrial Average is at 10,000 points on March 29.
1999	Human population size at 6 billion.
1999	Human Chromosome 22 DNA and genes are sequenced.
2000	Atmospheric CO_2 concentration has grown to '360 microliter/liter.
2000	Most distant manmade object. December 18, 2000, Voyager 1 spacecraft.
2001	NEAR spacecraft is first manmade object to land on an asteroid.
2001	Events from *2001: A Space Odyssey* begin.
2003	End of the World predicted by aliens called Zetas.
2003	Cassini spacecraft arrives at Saturn.
2003	Mars at opposition. 34.6 million miles. Closest to Earth in last 3,000 years.
2004	*Alderson Drive is perfected and allows interstellar travel (Jerry Pournelle).*
2004	Transit of Venus June 8.
2004	*Superconducting Supercollider creates a worm-hole bridge (John Cramer).*
2004	End of the World predicted by Arnie Stanton. Second Coming.
2005	*Eon. First expedition to hollowed-out asteroid uncovers The Way (Greg Bear).*
2006	End of the World.
2007	End of the World. Rapture occurs.
2009	*Flashforward. At the CERN something goes horribly awry.*
2010	World's oil supply will reach its maximum production and midpoint of depletion.
2010	Half of all automobiles and trucks will be electric.
2012	End of the World. Bible Code shows earth hit by devastating comet.

Time	Event and Comments
2012	Great Cycle of the Maya is 5,125 years in length, comes to an end.
2012	End of the World. Earth enters Photon Belt from the Pleiades.
2014	End of the World. Pope Leo XIX prediction ca. 1514.
2014	Planned launch of Terrestrial Planet Finder by NASA.
2015	Fossil industry will start to feel the cumulative heat of dwindling oil supply.
2015	Greenhouse gas emissions in developing world exceed industrialized world.
2016	End of the World. Biological plague released that kills all of humanity.
2016	Possible date for the fly-by of the New Horizons mission with Pluto.
2017	End of the World According to Sword of God Brotherhood.
2018	End of the World. Comet or Asteroid collides with Earth.
2019	*Cities in Flight. Gravitron polarity generator invented (James Blish).*
2020	About one-fourth of children will live in poverty.
2020	Skill shortage of nine million college-educated workers.
2020	Estimated world population could be at 8 billion people.
2020	Global natural gas production expected to peak by now.
2020	Massive and uncontrolled deforestation in Africa.
2020	U.S. Government will owe Social Security Trust Fund $3 trillion.
2020	Minority students will represent 46% of all students in United States.
2020	Mt. Kilimanjaro, complete loss of snow.
2020	*Podkayne of Mars. Intra-planet travel is commonplace. (Robert Heinlein).*
2022	*Soylent Green* events begin.
2024	End of the World. Asteroid hits Earth.
2025	*Childhood's End. 50 years after the Overlords arrived in 1975 (Arthur C. Clarke).*
2025	Light pollution triumphs. Milky Way invisible in eastern half of the United States.
2025	Two out of every three people on Earth will live in water-stressed conditions.
2028	End of the World. Asteroid 1997 XF11 within 600,000 miles. Could collide.
2030	End of the World. Predicted by Sun Microsystems chief scientist.
2030	The projected insolvency date for Medicare Trust Fund.
2030	One in four American women will be over age of 65.
2030	Peak aging of baby-boom generation. Total population will reach 50.8 million.
2030	The world's automobile population will surpass one billion.
2030	Global warming could flood New York subways, airports, coastal areas.
2030	World population 9.6 billion. 96% of addition in developing nations.
2030	*The Star Dwellers. Haertel faster-than-light drive invented (James Blish).*
2036	*The Light of Other Days. WormCam invented (Arthur C. Clarke).*
2038	End of the World. Massive starvation and polar icecaps melt.
2040	No distinction between a commercial airliner and a commercial launch vehicle.
2040	Population of California will double to 60 million people.
2042	*Timemaster. First expedition to Alpha Centauri. Warpgate (Robert Forward).*
2047	End of the World. Predicted by the Church of BLAIR.
2049	Robots will be doing many household chores such as cooking, lawn-mowing.
2050	The impact of global warming is expected to be obvious by now.
2050	People of color will comprise nearly 50 percent of the U.S. population.
2050	U.S. Glacier National Park will contain no glaciers.

Time	Event and Comments
2050	Jet contrails to be significant climate factor.
2050	Ozone layer will begin to repair itself.
2050	The world population may reach 10 billion.
2050	World energy consumption will triple by 2050.
2050	Photovoltaic solar energy: 18% of all electricity generation.
2050	Fuel-cell electric power will predominate.
2050	Petroleum reserves are finally exhausted.
2050	Death of the oil economy.
2050	Earth-dwellers will probably be able to travel to the Moon.
2050	Globe will warm up by 1.5–4.5 degrees Celsius if CO_2 reaches 600 ppm.
2053	*Third World War. Zefram Cochrane invents space warp drive (Star Trek).*
2058	*Lost in Space movie. Jupiter 2 is launched from Earth.*
2060	Population of the United States is now at 560 million Americans.
2061	*Babylon 5. First permanent lunar colony established.*
2061	Halley's Comet returns. This is 31st time.
2065	*Procyon's Promise. A Maker ship enters solar system (Michael McCollum).*
2076	End of the World. Bede the Venerable, 8th-century Christian theologian.
2083	*Alien movie begins its history.*
2084	Transit of Earth as seen from Mars on November 10.
2088	Computers will be twice as smart and insightful as any human being.
2090	Great Lakes water levels will likely be one to three feet lower.
2095	Chinese economy has grown by a factor of more than 50 since 2000.
2100	Two-thirds of the world's 6,700 languages will have become extinct.
2100	Average temperature increases worldwide by more than 3 degrees Celsius.
2100	Population 10 to 11 billion.
2100	Sea level in 2100, would be about 0.4 meters higher than in 2000.
2100	Atmospheric CO_2 increases to 700 ppm. Temperature 3.5° F warmer.
2100	*The Ring of Charon. Earth is accidentally destroyed (Roger MacBride Allen).*
2101	*Babylon 5. First Martian base established.*
2102	Polaris reaches its closest position with respect to the North Celestial Pole.
2119	*Orphans of the Sky. First Proxima Centauri Expedition (Robert Heinlein).*
2150	*Dr. Who and his traveling companions use his time machine and visit Earth.*
2115	*Childhood's End. All adult humans are now extinct (Arthur C. Clarke).*
2125	*Childhood's End. The Earth is destroyed (Arthur C. Clarke).*
2130	*Rama Revealed. Alien artifact enters the solar system (Arthur C. Clarke).*
2197	*The Engines of God. The Iapetus Expedition (Jack McDevitt).*
2199	*The Matrix. The world is now being run by an artificial intelligence.*
2200	*Galaxies Like Grains of Sand. The Procyon Colony is settled (Brian Aldis).*
2218	*Star Trek. First contact with Klingon Empire.*
2245	*Babylon 5. The Earth-Minbari War begins with the death of Dukat.*
2261	*Babylon 5. The Shadow-Vorlon war at Coridian.*
2264	*Star Trek. Captain Kirk begins the historic Five-Year Mission.*
2267	*Babylon 5. Drakh release plague on Earth. Humanity annihilated except Rangers.*
2300	*The End of Eternity. Humans discover the Temporal Field (Isaac Asimov).*

Time	Event and Comments
2341	*The Man-Kazin War begins (Larry Niven).*
2367	*Star Trek Next Generation. Borg attack Federation.*
2375	*Cities in Flight. Earth cities launched into space wander the galaxy (James Blish).*
2400	*Eon. Humans hollow out the asteroid Ceres and install The Way (Greg Baer).*
2439	*The Sundered Worlds. End of the universe (Michael Moorcock).*
2487	*Buck Rogers TV series.*
2512	*Antares Dawn (Michael McCollum).*
2525	Zager and Evans: This year begins their song.
2600	*The End of Eternity. Time travel created (Isaac Asimov).*
3001	*Ring. Space travel with GUT-drive and wormhole technology (Stephen Baxter).*
3100	Earth's next little ice age could occur.
3111	*Cities in Flight. New York is launched into space (James Blish).*
3175	*A Canticle for Leibowitz. Earth destroyed by nuclear war (Walter Miller).*
3262	*Babylon 5. On a decimated Earth, monks collect artifacts to reconstruct past.*
3500	*The Songs of Distant Earth. The quantum drive is invented (Arthur C. Clarke).*
3620	*Songs of Distant Earth. Earth destroyed as the Sun goes nova (Arthur C. Clarke).*
3850	*The Corridors of Time. Time travel is invented (Poul Anderson).*
4000	*The Chaos Weapon. Trans-continuum drives (Colin Kapp).*
4000	Earth's magnetic field vanishes, perhaps reversing polarity.
4004	*Cities in Flight. Rebirth of the universe (James Blish).*
4874	*Ring. Conquest of human planets by Squeem (Stephen Baxter).*
5088	*Ring. Conquest of human planets by Qax (Stephen Baxter).*
5500	*Ring. Third human expansion begins into galaxy (Stephen Baxter).*
10,000	*Ring. Humans become dominant sub-Xeelee species in galaxy (Stephen Baxter).*
11,500	Because of precession, Earth will then be closer to the Sun in July.
12,000	*The Wizard of Linn (A. E. van Vogt).*
16,000	Vega is now the North Pole Star.
20,000	Global climate would be heading toward another ice age.
23,000	A full-scale glaciation, much like the last one, will culminate.
23,125	The star Thuban is now the North Pole Star.
26,000	*Hari Sheldon establishes the Foundation (Isaac Asimov).*
26,390	*Dune. House Atreides moves to Arakis (Frank Herbert).*
27,827	Polaris returns to its closest offset from the North Celestial Pole.
30,000	Pioneer 10 passes close to nearby star Ross 248, 10 light-years distant.
50,000	Solar system enters a piece of "Local Fluff." Possible climate effects.
80,000	Peak of the next ice age.
82,000	*Galaxies Like Grains of Sand. Space travel resumes (Brian Aldis).*
100,000	*Ring. Human assaults on Xeelee concentrations begins (Stephen Baxter).*
202,000	*Return to the Stars. The Empire battles the Dark Worlds (Edmond Hamilton).*
657,208	*The Time Ships. The Morlock civilization (Stephen Baxter).*
802,701	*In the Time Machine chronology the Morlocks and Eloi saga unfolds.*
1 million	Gleise 710 reaches closest point to Sun at 1 light-years. Comet storms?
1 million	*Babylon 5. Sun goes nova. Rangers become Vorlon-like beings.*
1 million	*Ring. Xeelee final defeat of humans (Stephen Baxter).*

Time	Event and Comments
1 million	*City and the Stars. Humans colonized the galaxy (Arthur C. Clarke).*
3 million	*City and the Stars. Invaders enter Galaxy. Defeat humans (Arthur C. Clarke).*
3 million	*The End of Eternity. Humans will have evolved (Isaac Asimov).*
3 million	Betelgeuse Supernova by now.
4 million	*Ring. Destruction of Ring by photino birds (Stephen Baxter).*
5 million	*Ring. Last humans return to solar system in GUT ship (Stephen Baxter).*
5 million	Sun reaches the edge of the Local Bubble. Possible climate effects.
7 million	*The End of Eternity. The mysterious Hidden Centuries (Isaac Asimov).*
10 million	*Ring. Extinction of baryonic life (Stephen Baxter).*
15 million	*The End of Eternity. Humanity has completely died out (Isaac Asimov).*
20 million	Limit to numerical integration of dynamics of solar system.
20 million	35 supernovae in Sco-Cen OB association light up the sky.
40 million	*Galaxies Like Grains of Sand. The end of the Dark Millennia (Brian Aldis).*
66 million	Sirius leaves the main sequence and starts life as red giant.
100 million	*City and the Stars. Humanity confined to a few cities: Diaspar (Arthur C. Clarke).*
100 million	Solar luminosity 1% higher. Global temperature is 10 degrees C higher.
100 million	Saturn's rings disappear.
200 million	Global temperature will be 20 degrees C higher than in 2000.
225 million	Sirius is a mature red giant with a red color and a magnitude of −5.0.
250 million	Continent distribution will once again be like ancient Pangaea.
300 million	The binary pulsar PSR1913+16 collapses in a spectacular supernova.
300 million	Mean global temperature 30 C higher.
342 million	Sirius enters a planetary nebula phase.
400 million	Mean global temperature 40 C higher.
500 million	Atmosphere so short of carbon dioxide that all plants die.
500 million	Global temperature 50 C higher.
600 million	The Sun will be 5% brighter than it is today.
600 million	Earth temperature will have risen to about 350 K or 80 C.
750 million	Sagittarius dwarf galaxy eaten by Milky Way.
1 billion	*The City and the Stars. Human interstellar civilization (Arthur C. Clarke).*
1 billion	*Galaxies Like Grains of Sand. Human extinction across galaxy (Brian Aldis).*
1 billion	Earth will lose its oceans and revert to its former lifeless conditions.
1.1 billion	Sun will be 10% brighter than today.
1.1 billion	Earth would start to lose its water via a "moist greenhouse."
1.3 billion	Surface temperature 100 C. The surface pressure of the atmosphere doubles.
1.5 billion	*The City and the Stars. Humans create incorporeal beings (Arthur C. Clarke).*
2 billion	Global temperature about 400 K or 130 C.
2 billion	Eccentricity of the orbit of Mars exceeds 0.2.
2 billion	*The City and the Stars. Alvin leaves Diaspar (Arthur C. Clarke).*
2 billion	Moon no longer helpful in stabilizing our rotation axis. Chaotic behavior eminent.
2.3 billion	The solar system makes its tenth orbit of the galactic center.
3 billion	Sun reaches its hottest surface temperature of 5843 on Main Sequence.
3 billion	Solar luminosity increases 33%, global temperature 450 K or 180 C.

Time	Event and Comments
3 billion	Andromeda Galaxy collides with Milky Way.
3.5 billion	Runaway greenhouse effect. The oceans will evaporate into space.
3.5 billion	Mercury suffers close encounter with Venus and is ejected from solar system.
3.5 billion	Earth now a twin to Venus but with water-vapor-rich 100-bar atmosphere.
4.8 billion	Earth's temperature will have risen to about 1600 K or 1330 C.
5.0 billion	Sun starts to become a red giant.
7.5 billion	*City at World's End. Middletown propelled into future (Edmond Hamilton).*
7.55 billion	Venus at 1 AU Earth 1.38 AU. Mercury engulfed, but Venus just escapes.
7.68 billion	Sun engulfs Venus and Earth. Mars spared at a distance of 2.25 AU.
11 billion	Youngest globular clusters have now evaporated away.
17 billion	*The City and the Stars. Mad Mind battles Vanamonde (Arthur C. Clarke).*
50 billion	Earthlings will only see galaxies that are now our closest neighbors.
100 billion	Universe will be a dark and lonely place.
100 billion	The universe is likely to remain hospitable to life.
1000 billion	Stars die. No more fusion-powered stars. Galaxies filled with corpses.
1500 trillion	*Ring. Stars evaporate from galaxies (Stephen Baxter).*
10^{17}	Passing stars rip planets from stellar corpses.
10^{18}	Galaxies dissolve.
10^{20}	Orbits of planets decay via gravitational radiation.
10^{30}	The Black Hole Age.
10^{31}	Universe filled with supermassive black holes.
10^{34}	All carbon-based lifeforms become extinct (due to lack of atoms).
10^{35}	Proton and neutron will decay if current GUT theories are correct.
10^{65}	Stellar-mass black holes evaporate!
10^{65}	Ordinary matter liquifies due to quantum tunneling.
10^{66}	Solar-mass black holes evaporate via Hawking process.
10^{97}	Galaxy-sized black holes will also evaporate.
10^{100}	Supermassive black holes evaporate.
10^{110}	The Dark Age: that no known process will ever change things.
10^{122}	Protons decay via Hawking process.
10^{1500}	Ordinary matter surviving GUT or Hawking process decays into iron.
$10^{10^{26}}$	All iron nuclei collapse into black holes.

NOTE: Future predictions from various science fiction books, TV series, and movies are shown in italics. Other predictions are based on scientific sources. Some of these forecasts contradict other science-based predictions. Mya. = Million years ago; Kya = Thousand years ago. Scientific notation for far future expressed as 10^3 for 1000, 10^4 for 10000 and $10^{10^{26}}$ for ten raised to the 10^{26} power.

Table 9: Nearest Stellar Black Holes, Neutron Stars, and White Dwarfs

Name	Type	Distance (lys)	Comments
Sirius-B	WD	8.8	1.02 Msun
40 Eridani B	WD	9.45	0.43 Msun
Procyon-B	WD	11.4	0.63 Msun
Van Maanens Star	WD	13.7	DZ - helium and metal-rich
GL 432B	WD	31.9	DC-type
GL 288 B	WD	49.5	DC-type
WD0346+246	WD	91	DA-type: coolest known
GL 848.1C	WD	160.4	DC-type
Geminga	NS	320	340,000 years old
RX J185635-3754	NS	< 400	1.1 million years old
PSR B0950+08	NS	420	17 million years old
PSR J0108-1431	NS	423	160 million years old
PSR J0437-4715	NS	450	2 billion years old
V4641 Sgr	BH	1,600	8.7–11.7 Msun
PSR J1908+0734	NS	1,900	4 million years old
A0620-00	BH	2,700	7.4 Msun
Nova Muscae	BH	3,000	3 Msun
PSR B1259-63	NS	4,900	330,000 years old
Crab Pulsar	NS	6,300	950 years old
GRO J0422+32	BH	6,500	6 Msun
GS 2000+25	BH	6,500	8–10 Msun
Cygnus X-1	BH	7,500	6 Msun
V404 Cygni	BH	8,000	9 Msun
GRO 1655-40	BH	10,000	7 Msun
XTE J1550-564	BH	17,000	10.5 Msun
GX339-4	BH	18,000	5 Msun
LMC X-3	BH	160,000	> 3 Msun

NOTE: White dwarfs (WD) and neutron stars (NS) tend to have masses between 1 and 2 Msun. Neutron stars are more interesting because of their ages, which distinguish them. White dwarfs are distinguished by their element abundance classifications (e.g. DA, DC, and DZ).

Table 10: Table of Identified Supermassive Black Holes

Name	Discovery	Estimated Mass
Milky Way	2002	2.6 million
Centaurus-A	2001	200 million
M-31	1994	30 million
NGC 3115	1992	1 billion
M 32	1995	2 million
NGC 4594	1994	500 million
NGC 3377	1997	100 million
NGC 4258	1995	40 million
M 87	1994	3 billion
SDS 1044-1025	2001	>3 billion
NGC 4697	2000	50–100 million
NGC 4473	2000	50–100 million
NGC 821	2000	50–100 million
NGC 4261	1995	400 million
NGC 3379	1997	50 million
NGC 4486b	1997	500 million
Messier-15 (Glob C)	2002	3,900

NOTE: Discovery year in column 2, estimated mass in multiples of solar value in column 3.

Table 11: Travel Times to the Planets at Various Speeds

Method Speed =	Shuttle 28,000mph	Galileo 54,000	Ion A 65,000	Ion B 650,000	Solar Sail 200,000
Mercury	52d	27d	22d	2.2d	7.3d
Venus	100d	52d	43d	4.3d	14d
Mars	210d	109d	90d	9d	29d
Jupiter	1.9 yr	1 yr	303d	31d	100d
Saturn	3.6 yr	1.8 yr	1.5 yr	55d	179d
Uranus	7.3 yr	3.8 yr	3.1 yr	113d	1 yr
Neptune	11.4 yr	5.9 yr	4.9 yr	179d	1.6yr
Pluto	15.1 yr	7.8 yr	6.5yr	238d	2.1 yr

NOTE: Ion Drive using a constant thrust of A) 0.1 pounds B) 1 pound with turn-around deceleration added. Two year's acceleration to reach top speed. Solar Sail—Interstellar Probe estimated speed (solar wind = 450 km/sec or one million mph).

Table 12: Notable Astronomical Events in Recorded History

Date	Event	Impact
May 3, 1374 B.C.	S. Eclipse	Ugarit Eclipse—oldest documented sighting.
June 3, 1301	S. Eclipse	Early Chinese eclipse observation.
April 16, 1177	S. Eclipse	Homer's *Odyssey.*
June 15, 762	S. Eclipse	Assyrian eclipse.
April 6, 647	S. Eclipse	Archilochus eclipse.
May 28, 584	S. Eclipse	Heroditus eclipse.
October 2, 479	S. Eclipse	Xerxes eclipse.
August 28, 412	L. Eclipse	Siege of Syracuse.
April 15, 405	L. Eclipse	Fire in the Temple of Athena.
September 20, 330	L. Eclipse	Arbela battle of Alexander the Great.
240	Halley's C.	First sighting by Chinese astrologers.
April 17, 6	Conjunction	Star of Bethlehem in Leo/Ares.
March 23, 4	L. Eclipse	Death of Herod.
Sept 27, 14 A.D.	L. Eclipse	Death of Augustus.
33	L. Eclipse	Crucifixion of Jesus.
37	Aurora	Caesar sends soldiers north to put out fire.
60	Halley's C.	Nero has all possible successors executed.
March 20, 71	S. Eclipse	Plutarch's eclipse.
November 24, 569	S. Eclipse	Eclipse preceding birth of Muhammad.
616	Meteorite	China meteorite kills 10 solders in a camp.
May 5, 840	S. Eclipse	Treaty of Verdun.
902	M. Storm	Leonids recorded by Arab astronomers.
December 22, 968	S. Eclipse	Constantinople. Leo Diaconus discovers corona.
1006	Supernova	Lupus. 2 yrs. brighter than Venus.
December 28, 1047	Conjunction	M-J-S-Sun less than 3.4 deg.
1054	Supernova	Messier 1. m= -3.5. Chinese, Japanese, Pueblo art.
1066	Halley's C.	Battle of Hastings—Mass terror—Europe.
April 24, 1146	Conjunction	M-J-S-Sun less than 3.2 deg.
1181	Supernova	Cassiopeia. 6 months. m= -1.0.
September 20, 1186	Conjunction	M-J-S-Sun less than 7 deg.
January 15, 1192	Aurora	Seen in Flanders.
May 14, 1230	S. Eclipse	Major European eclipse.
January 3, 1285	Conjunction	M-J-S-Sun less than 9.5 deg.
1341	Meteorite	China meteorite shower kills people.
June 21, 1385	Conjunction	M-J-S-Sun less than 3.8 deg.
November 9, 1425	Conjunction	M-J-S-Sun less than 8.4 deg.
May 22, 1453	L. Eclipse	The fall of Constantinople.

Date	Event	Impact
1456	Halley's C.	Blamed for earthquakes, illness.
1490	Meteorite	Ch'ing-yang Meteorite Shower kills thousands.
March 1, 1504	L. Eclipse	Christopher Columbus eclipse in Americas.
September 1511	Meteorite	Monk killed by meteorite fall.
February 18, 1524	Conjunction	M-J-S-Sun less than 7 deg.
1531	Halley's C.	Pisarro conquest of Incas.
October 1572	Supernova	Tycho's SN. −4.0. Cassiopeia, lasted 483 days.
1577	Comet	Overturned Aristotelean crystalline spheres.
October 5, 1591	Aurora	Seen in Europe.
1599	Comet	Montezuma sees Cortez as fulfillment of prophecy.
1604	Supernova	Ophiuchus. 365 days. m= -2.6.
1607	Halley's C.	Observed by Johannes Kepler.
July 9, 1622	Conjunction	M-J-S-Sun less than 7.4 deg.
August 22, 1624	Conjunction	M-J-S-Sun less than 9 deg.
November 17, 1630	M. Storm	Leonids light up sky for Kepler's funeral on 11/15.
1639	Venus T.	First transit of Venus across the Sun observed by humans.
1639	Meteorite	Ch'ang-shou county meteorite kills 10s of people.
1650	Meteorite	Monk in Milan killed by meteorite. Artery severed.
1658	Supernova	Cass-A. (3 Cass? seen by Flamsteed before SN).
1680	Comet	The Great Comet.
February 10, 1681	Aurora	Seen in Europe.
1682	Halley's C.	Orbit calculated by Edmund Halley.
May 3, 1715	S. Eclipse	Edmund Halley's eclipse.
March 17, 1716	Aurora	Seen in Europe.
1729	Comet	Great Comet of 1729.
1759	Halley's C.	First prediction by Edmund Halley.
1744	Comet	Spectacular and famous comet de Chesaux's Comet.
June 6, 1761	Venus T.	Major international observing campaign.
August 5, 1766	S. Eclipse	Captain Cook's eclipse.
June 3, 1769	Venus T.	Captain Cook expeditions.
1799	M. Storm	von Humbolt mentions intense storm.
April 26, 1803	Meteorite	Major fall on L'Aigle France.
January 15, 1805	L. Eclipse	Lewis and Clark Expedition eclipse.
1811	Comet	Great Comet of 1811.
November 12, 1833	M. Storm	Major Leonid shower terrifies millions.
1835	Halley's C.	Blamed for Alamo massacre and NY fire.
September 15, 1839	Aurora	Great Aurora seen in England.
1843	Comet	Great Comet of 1843.
November 17, 1848	Aurora	Great Aurora. California Cuba, Europe. Telegraphs.

Date	Event	Impact
August 28, 1859	Aurora	Great Aurora. Cuba, Rome, Europe.
1858	Comet	Donati's Comet.
September 2, 1859	Aurora	Athens. San Salvador. Major white light flare.
September 2, 1861	Conjunction	M-J-S-Sun less than 3.6 deg.
1861	Comet	Great Comet of 1861.
February 4, 1872	Aurora	Great Aurora. Mexico, Athens, India.
December 9 1874	Venus T.	Venus Transit. Major international expeditions.
January 22, 1879	S. Eclipse	Zulu War Eclipse.
1881	Comet	Great Comet of 1881.
November 18, 1882	Aurora	Great Aurora. Cuba, Mexico, US. Telegraph outages.
1882	Comet	Great Comet. Photographed for first time.
December 6, 1882	Venus T.	Photos and major public interest.
March 30, 1886	Aurora	Great Aurora. England, China, Japan, India.
February 13, 1892	Aurora	Great Aurora. Iowa, New York.
March 30, 1894	Aurora	England.
1897	Meteorite	Meteorite kills horse. New Martinsville, WV.
September 9, 1898	Aurora	Great Aurora. Omaha, Tenn., New York.
1899	M. Storm	No Leonids. Public disappointed in astronomers.
November 1, 1903	Aurora	Great Aurora. Europe, N. America. France telegraphs.
1907	Meteorite	Hsin-p'ai-wei Meteorite kills Wan family.
June 30, 1908	Meteorite	Tunguska explosion.
September 25, 1909	Aurora	Great Aurora. Singapore.
1910	Halley's C.	Comet Hysteria. Suicides. Mayhem. Deadly gas.
January 1910	Comet	Great Daylight Comet of 1910.
April 17, 1912	S. Eclipse	The *Titanic* eclipse.
April 25, 1915	Meteorite	China. Meteorite rips woman's arm off.
July 4, 1917	L. Eclipse	Lawrence of Arabia's eclipse.
August 7, 1917	Aurora	Great Aurora. New York. Illinois.
May 29, 1919	S. Eclipse	Einstein's eclipse.
March 22, 1920	Aurora	Great Aurora. Boston, Washington, Norway.
May 14, 1921	Aurora	Great Aurora. England, Samoa, Jamaica.
January 24, 1925	S. Eclipse	NY City's Winter Morning eclipse.
January 26, 1926	Aurora	Great Aurora. U.S., Scandinavia. London blackout.
December 8, 1929	Meteorite	Meteorite kills a person in Yugoslavia.
August 31, 1932	S. Eclipse	Great Maine eclipse.
January 25, 1938	Aurora	Great Aurora. U.S., North Africa. Fatima Aurora.
March 24, 1940	Aurora	Great Aurora. N. Dakota. Blackout in northeast.
September 18, 1941	Aurora	Great Aurora. TV/Radio disruption in N. America.
July 26, 1946	Aurora	Great Aurora. Boston, Tenn.
1948	Meteor	Nebraska Fireball. 2,300-pound meteor recovered.

Date	Event	Impact
November 6, 1951	Meteorite	Bremerton, Wash. Man injured by meteorite.
1954	Meteorite	Meteorite injures woman. Sylacauga, Alabama.
February 24, 1956	Aurora	Iceland, Alaska. Intense cosmic ray blast.
September 13, 1957	Aurora	Great Aurora. Mexico.
February 10, 1958	Aurora	Great Aurora. N. America, USSR. SW Blackout.
October 1965	Comet	Ikeya-Seki.
November 17, 1966	M. Storm	Major Leonid shower over central U.S.
February 8, 1969	Meteorite	Allende fall. Famous and spectacular.
March 23, 1969	Aurora	North America.
September 28, 1969	Meteorite	Murchison fall. Spectacular and famous.
March 1970	Comet	Comet Bennett—very bright.
August 4, 1972	Aurora	Great Aurora. N. America, Canada, Scandinavia.
August 10, 1972	Fireball	Daytime over Grand Tetons—1,000 tons.
January 1974	Comet	Kohoutek—a big but famous fizzle.
March 1976	Comet	Comet West.
March 5, 1981	Aurora	Great Aurora. Colorado.
February 1987	Supernova	Large Magellanic Clouds. Optically discovered.
March 13, 1989	Aurora	Great Aurora. Canada, US. Quebec power blackout.
October 9, 1992	Meteorite	Peekskill, NY, meteorite hits car.
July 16, 1994	Comet	Shoemaker-Levy collides with Jupiter.
March 26, 1996	Comet	Hyakutake—Very bright.
January 11, 1997	S. Storm	Telstar 401 satellite damaged.
April, 1997	Comet	Hale-Bopp. Much celebrated.
May 5, 2000	Conjunction	M-J-S-Sun less than 18 degrees.
October 14, 2001	Meteorite	Canadian Fireball.
July 15, 2000	S. Storm	Bastille Day Storm.
April 2, 2001	S. Storm	Major X20 solar flare.
March 2003	Meteorite	Several Illinois towns hit by debris. Damage.
June 8, 2004	Venus T.	First observation in 122 years.

NOTE: Abbreviations—Venus T. (Venus Transit of Sun), S. Eclipse (Total Solar Eclipse), S. Storm (Solar Storm), Halley's C. (Halley's Comet), L. Eclipse (Lunar Eclipse), M. Storm (Intense Meteor Shower, Meteor Storm).

Index